Energy
and the future

Text and illustrations by
David A. Hardy FRAS, AFBIS

Energy
and the future

World's Work Ltd

Also by David A. Hardy

Air and Weather
Light and Sight
The Solar System
Rockets and Satellites

Text and illustrations copyright © 1979 by David A. Hardy
Published by World's Work Ltd
The Windmill Press, Kingswood, Tadworth, Surrey
Reproduced by Graphic Affairs Ltd, Southend
Made and printed in Great Britain by
William Clowes & Sons Limited, Beccles and London
SBN 437 45199 2

To Ruth

Contents

Introduction	8
The many faces of energy	10
All change	13
External and internal energy	14
Weights and measures	17
More units	18
Heat and motion	20
Caloric	21
Motion carried	22
A matter of elements	23
Mixed reactions	24
Building an atom	27
The Periodic Table	28
Attraction of opposites	30
Making the sparks fly	31
The current flows	33
Names to remember	34
Putting electricity to work	36
Let there be light	38
Making the quantum jump	40
A new light on atoms	42
Enter radioactivity	44
Splitting atoms	45
War on the nucleus	48
Atoms for war and peace	49
Another energy crisis	52
The burning rock	54

Coal for the future	56
From coal to gas	58
Where the gas goes	60
When will the gas go out?	62
Black gold	64
The costly quest	66
Unconventional oil	66
The nuclear alternative	70
Problems, problems. . . .	72
Fishing for fusion	74
Electricity for all	76
Are we cool?	79
Powerhouse in the sky	80
Energy on the beam	81
Earth power	84
Sea power	86
Gravity power	88
Moon power	90
Air power	92
Plant power	94
And back to gas	96
The ideas keep coming	98
Crisis – what crisis?	100
Epilogue	104
Selected bibliography	106
Further information	107
Index	109

Introduction

The so-called oil crisis of 1973 brought to light a remarkable number of research projects aimed at replacing our diminishing supplies of fossil fuels as energy sources. Many of these, though promising, were relatively small and privately-funded. Had the clear warning signs given out by the Oil Producing Exporting Countries in 1973 been heeded by governments and the public, there would have been no need to be caught napping six years later. OPEC are unlikely to increase present production ceilings since oil in the ground is worth more to them than extra dollars they cannot use. Even without the 1973 Middle East war, the annual increase of 7.5% in world oil consumption could not have continued for more than an extra five to ten years before demand exceeded supply; so our most precious resource is really time.

In 1977 I set out to review all the alternative energy sources which could be available when the oil finally stops flowing – or preferably before. In ploughing through the reams of reports, documents and publications which this research entailed, I discovered what a large amount of prior knowledge was often expected from the reader, even in a 'non-technical' article. Does everyone know how much oil is in a 'barrel'? More basically, how did all that oil (and coal) get into the ground, and how is it removed? Who first did so? What is 'light water' – or 'heavy water'? By what strange magic does nuclear energy get into our house wiring as electricity, and is it really dangerous? For that matter, who invented it, and how did they know that atoms exist at all? What *is* an atom. . .?

Every avenue of enquiry led to another, even more fascinating. In this book I have tried to tell the whole story of energy in its many forms, from the beginning: who discovered what, how and when, and how it was eventually put to practical use – or may be. I also came upon many terms, units and symbols. Having been educated, like many people, with older systems, it seemed a good idea to devote a few pages to the modern versions which are becoming internationally accepted.

Above all, I tried to be completely impartial. Every new or radical method of energy production is advocated by someone (or a whole team) who has become expert in that field – but for each expert in one field there are ten in others who will claim that their own development is the one which should be the ultimate solution to mankind's energy problems. I have set them out, side by side, for the reader's consideration, along with more conventional or established sources. Technicalities and mathematics have been kept to an essential minimum.

In summing up on the final pages I have suggested what seems to me the most likely scenario, based on present trends, on the calculations and opinions of authorities in the field of energy as a whole, and the sort of policies which history shows can be expected from politicians. My personal views are restricted to the epilogue on page 104.

DAVID A. HARDY
July, 1979

Energy and the future

From Aristotle's classic theory of 'the four elements' to a Moon landing is a giant step indeed. The theory actually held up the progress of science until Galileo's experiments sparked off the scientific renaissance of the 17th century in Europe. Commander Scott's verification of Galileo's cannon ball demonstration was light-hearted, but the Apollo missions used virtually every form of energy described in this book at some stage.

The many faces of energy

We hear almost daily about the 'energy crisis' caused by our using up the world's energy reserves too fast, so that we shall be unable to meet future demands. These reserves consist mainly of *fossil fuels* – coal, oil and gas, which today produce almost 99% of our energy. But where did this energy come from in the first place? From the Sun, beating down on the forests and swamps which covered our planet hundreds of millions of years ago. Plants have the unique ability to extract and store energy from sunlight; we can release this energy when we eat plants or burn wood or vegetation, which man began to do as soon as he discovered fire.

The energy of the Sun has many other guises, once it reaches Earth, as we shall see later. But it may come as a surprise to realise that even the power of our own muscles comes, via a long chain of events, from that bright object in the sky. The story of how Man gradually learnt about the various forms of energy, how he measures them, and how theories were proved (or discarded) and finally put into practical use, is a fascinating one.

A caveman used only his own energy, derived from eating plants or meat from animals which in turn ate vegetation, which stored energy from sunlight.... He may have used his strength to rub two sticks together with sufficient friction to produce heat and then fire to cook his food and provide warmth and light at night. Muscles, food, plants, sunlight, friction, heat, light: can these *all* be energy? What *is* energy? Whereas *matter*, the other constituent of the universe we live in, can be seen, felt, smelled, held or weighed, energy seems intangible, almost abstract.

Ancient philosophers, and especially the Greeks, often wondered about this. Around 500 BC Heraclitus believed that fire could be transformed into water and water into earth, and vice versa; and this search for a 'unifying principle' to link together the observable phenomena of the universe continued with Democritus, Plato and Aristotle. According to Aristotle, a strange 'Unmoved Mover' worked tirelessly to keep the universe in motion. This notion that everything that moved needed a continual force – perhaps from the air rushing in behind, it was suggested – to keep it in motion persisted for hundreds of years. The theory does not seem to explain why a thrown stone eventually falls to earth, but it was believed that the speed of a falling object depended upon its weight – a heavy body falling faster than a light one.

It was the great Italian scientist Galileo Galilei who proved this to be a fallacy, and gave us the first true insight into the 'mechanics' of energy. There is a story that he pushed two (or three) iron balls, weighing from 1.5 to 45 kg, from the top of the 55 metre-high Tower of Pisa, finding that they fell and hit the ground together. (That *all* bodies do fall at the same rate in the absence of air resistance was shown dramatically by Astronaut David Scott when, during the Apollo 15 mission, he dropped a hammer and a falcon feather on the Moon.) Towards the end of the

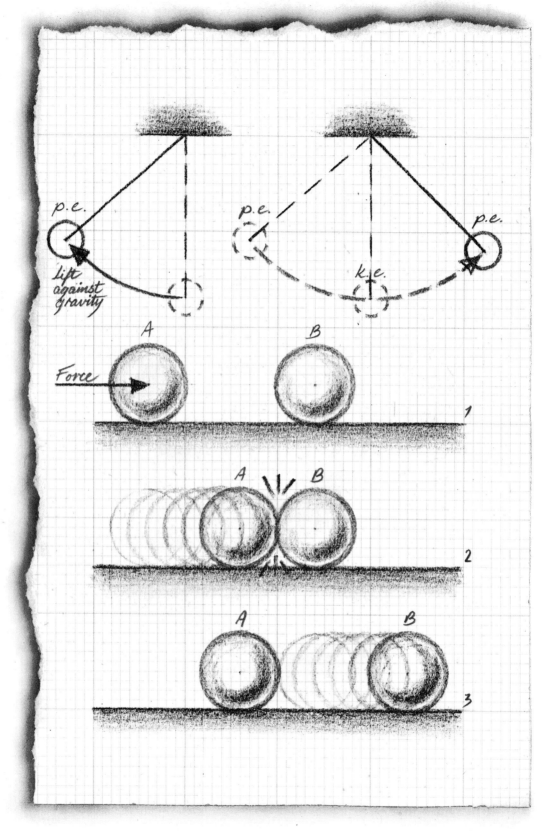

(Top) Potential and kinetic energy, as in a pendulum or swing. The weight can never rise quite as high as it started, due to air friction. The swing is exaggerated in the diagram. *(Below)* When two marbles collide one may be stopped but the other moves off, as Huygens found. Eventually friction takes its toll.

16th century Galileo also experimented with pendulums after watching a swinging lamp in the cathedral at Pisa, timing it with his own pulse. He found that even though the arc of swing grew a little shorter each time, the time taken from side to side remained constant.

All change

By now you may be asking 'what has all this to do with the sort of energy we use every day?' The point is that Galileo's experiments, followed in the 17th century by the studies of mathematicians such as Descartes in France (and later Holland), Leibniz in Germany, Newton in England, and Huygens in Holland, gradually paved the way to the *Law of Conservation of Energy*, proposed by the German physicist Hermann von Helmholtz in a book published in 1847. This states that 'energy cannot be created or destroyed, only changed into different forms'.

For instance, Gottfried Wilhelm Leibniz, repeating Galileo's studies with falling bodies, realised that although they hit the ground at the same time, a large body would strike with more *force* than a small one – and do more damage. Similarly, the force of the impact depends upon how far the object falls. So Leibniz was able to say that force could be measured by finding the weight of a body and its speed when it is brought to rest. He called the impact *vis viva* ('life force'); the ability of an object to do damage to another object. Today we would use the term 'kinetic energy' to describe the same quality.

Kinetic energy (k.e.) is the energy a body has due to its motion. A hammer head does work by overcoming a force when it strikes something – preferably a nail. A heavy flywheel stores energy in the form of *rotational* kinetic energy, so keeps an engine running smoothly between piston strokes. A moving 2 kg object has twice the kinetic energy of a 1 kg object moving at the same speed; but doubling the speed of an object *quadruples* its energy. (In technical language: 'the kinetic energy of a moving body is directly proportional to its mass and proportional to the square of its speed'. Note that the term *mass* is used here, rather than 'weight'; the mass of an object remains the same whether it is on Earth, the Moon, or anywhere else in the universe, because it is a measure of the amount of material in that object. Weight, on the other hand, is governed by the pull of gravity on that mass. So, strictly speaking, the word 'weight' should read 'mass' in Leibniz's experiments, described earlier.)

Christian Huygens showed in 1699 that when two moving objects (say, two marbles) collide, the sum of their vis vivas is the same before and after the collision; one may be slowed, but the other will speed up. In other words, no kinetic energy

Energy and the future

A small piece of rock, drawn into Earth's atmosphere by gravity, has much potential energy. Friction with air particles causes it to glow red- then white-hot until (if large enough) it hits the ground and expends most of its remaining energy by blasting a crater, doing 'work' by moving earth. It then cools, by radiation, conduction into the material around it, and convection in the air. *Inset* 1. At normal temperatures the molecules in a body remain firmly in place. 2. When heated they begin to move around and often collide.

is lost, it is merely changed.

Galileo's cannon ball at the top of the Tower of Pisa possessed *potential energy* (p.e.). This is the energy something has due to its position or state. When anything is lifted vertically, work is done against its weight; this is stored in the object as *gravitational* potential energy. A Frenchman, Lazare Carnot, realised this in 1803 and named it 'latent' vis viva, over two centuries after Galileo's experiments. The energy in the swinging pendulum is either k.e. or p.e. or a mixture of both. At the extreme end of each swing it is all p.e. As it passes through its lowest point it is all k.e. In between, it is partly k.e. and partly p.e. Similarly, as our cannon ball fell and its speed increased, it gained in k.e. at the expense of its p.e., the loss of one being exactly equal to the gain in the other (if we ignore the energy given to air particles as they were moved out of the way), as one would expect from the Law of Conservation of Energy. The energy to raise the cannon ball against gravity in the first place came from chemical changes in Galileo's muscles, which in turn came from the food he had eaten . . . you know the rest.

External and internal energy

Isaac Newton was able to formulate his famous three Laws of Motion by linking together the principles of motion in objects (including the heavenly bodies), which he published in 1687 in a great book known as the *Principia*. The term 'newton' is now used as an absolute unit of force, as we will see later.

First, let's go back to our falling cannon ball. What happens to its potential and kinetic energy once it reaches the ground? The simplest definition of 'energy' is 'the ability to do *work*' – work being said to be done when anything is moved or altered. Certainly anything in the way of the cannon ball would be 'moved or altered', so the ball can be said to have done work. But also, upon hitting the ground a little *heat* would be generated, in both ball and earth or whatever.

So now we have seen the kinetic energy of motion transformed into another form of energy altogether – heat energy. Although we ignored the friction of air particles on the falling ball, it is far from negligible where higher speeds are concerned. Much of the energy expended in sending the Apollo spacecraft to the Moon was given out again as the Command Module plunged into Earth's atmosphere on its return – so much so that special layers of the module were burned away in a blazing trail of flame. Our caveman performed the same transformation when he rubbed two sticks together to make fire.

The p.e. and k.e. of a falling object are sometimes described as *macroscopic*, meaning that its effects are visible to the naked eye. It also contains another kind of

Meteorite

Red hot

Atmosphere

White hot
– vaporizes

Convection current

Radiation

Crater

Conduction

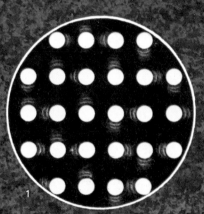
1

2

Energy and the future

Just a few of the scientists who led the way to an understanding of energy. Newton is best known for his work with gravity, motion and light; Huygens, whose 'vis viva' became kinetic energy, later opposed Newton's particle theory of light. Watt, as is well known, improved Newcomen's steam engine and enabled the energy of steam to be harnessed. Boyle actually financed publication of Newton's *Principia*; he also demolished the Aristotelian 'four elements' theory.

Sir Isaac Newton 1642–1727

Christan Huygens 1629–1695

James Watt 1736–1819

Robert Boyle 1627–1691

energy: the *internal* energy of the *molecules* inside, which may be called *microscopic*. The molecules which make up any substance are in continual motion; this internal energy increases as the temperature of the object rises. So as the ball hits the ground its energy is partly converted into internal energy, and its molecules – and those of the ground and air around it – speed up a little. This excess heat, given out by *radiation*, *conduction* and *convection*, as in the diagram on page 15, is eventually absorbed by the surroundings and dispersed. This 'degradation' is the ultimate fate of all energy; it may be that the 'end of the universe' will occur when all the energy in the universe has been distributed uniformly as internal molecular energy at the same temperature.

The definition of energy as the capacity to do work came originally from the English physicist, Dr. Thomas Young, who also coined the actual word 'energy' in 1807, from a Greek word meaning 'work'. This replaced the old term 'vis viva'.

Weights and measures

As soon as they have identified anything, scientists want to be able to measure it. A more accurate definition of work is 'the application of a force through a distance', and during the Industrial Revolution in Britain a system of units of measurement was introduced, based on the amount of work done when a weight of one pound is raised a distance of one foot: the *foot-pound*. This is still in use in some places, but international methods of standardising the units used in measuring *all* forms of energy are taking over, as below.

The *rate* at which work is done is known as *power*, so a machine may be said to have a power-output measured in various ways, one of the earliest of which was *horsepower* (h.p.) – a term used by James Watt around 1770 to describe the power of his newly-developed steam engine. He estimated that a horse could raise 33,000 pounds (as a round figure) to a height of one foot by pulling for one minute at this weight suspended by a pulley. This could also be written as '550 foot-pounds per second'. However, James Watt's name is perhaps better remembered by the unit named after him in the modern *SI* (*Système Internationale d'Unites*) system – the watt.

Until quite recently a system of units based on the centimetre, gramme and second (CGS) was widely used. In this the unit of work or energy is called the *erg*, defined as: 'the work done (or the energy expended) when a force of one dyne is exerted through a distance of one centimetre'. A *dyne* (dyn) is: 'the force required to accelerate one gramme, one centimetre per second per second'. This may take a little thinking about, as it involves a new term, *acceleration*. We all know the

Energy and the future

acceleration of a car; it is easy to see that if it accelerates (that is, its speed increases) at the rate of, say, one metre per second *every* second, it will soon be going pretty fast!

Galileo's cannon ball was accelerated by gravity as it fell, at the rate of about 981 centimetres per second per second. It did not have very far to fall, but the returning Apollo module, accelerating freely towards the Earth at the same rate, hit the atmosphere at some 11 kps, or 40,000 kph. The symbol for acceleration due to gravity is an italic 'g' (to distinguish it from 'g' for grammes), and it can be positive or negative. We normally call negative acceleration, or slowing down, 'deceleration'.

However, I have included the CGS units mainly because you may come across them in older books. In this book I use SI units, which are based on the MKS (metre, kilogramme, second) system. In the SI system the unit of power is the *watt* (W), as mentioned above: this is 'a rate of working of one joule per second'. A *joule* (J) is the SI unit of work: 'the work done when a force of one newton moves through one metre (in the direction of the force)'. And a *newton* (N) is: 'the force required to accelerate one kilogramme, one metre per second per second'. If you look again at the CGS units above you will see how the newton compares with the dyne – it differs only in the units of mass and length. A watt is equivalent to only 0.00134 h.p.

More units

We already know the origin of the term 'newton'. The joule was named after an English brewery owner, James Prescott Joule, who started experimenting around 1840 with rotating paddles in an insulated container of water, in order to heat the water by mechanical energy. The paddles were driven by the p.e. of falling weights (as in a grandfather clock), and after many years of painstaking effort Joule stated that to raise the temperature of one pound of water by 1°F took 772 foot-pounds. The modern figure is 778 foot-pounds. Joule's 'mechanical equivalent of heat', also known as a *British thermal unit* (Btu), was given the symbol italic J – not to be confused with J for joules of work – and was used for converting joules into calories or back.

Calories seem to be mainly the concern of 'slimmers' nowadays, but the term *calorie* (cal) in our context is defined as 'the quantity of heat required to raise the temperature of one gramme of water through 1°C'. A larger unit, the kilocalorie (kcal), sometimes called the Calorie, with a capital C, when it is used to measure the energy value of foods, is also used; 'one kilogramme' then replaces 'one gramme' in the above definition. If we convert the result of Joule's paddle experiment into SI units, we find that 1 calorie equals 4.2 joules, or 1 kilocalorie equals 4,200 joules.

In his attempts to prove that potential and kinetic energy could be converted into internal energy, Joule even measured the temperature of the water at the top and bottom of a Swiss waterfall, and found it *very* slightly warmer at the bottom.

Since many of these units are so small, the prefixes 'kilo-' and 'mega-' are commonly used, meaning a thousand and a million units respectively. We can also use a 'shorthand' method in figures by writing 1,000 as 10^3, 1,000,000 as 10^6 and so on – the upper figure representing the number of noughts (a 'minus' denotes a fraction). So we can speak of the *kilojoule* (kJ) for 10^3 J, and *megajoule* (MJ) for 10^6 J; and of the *kilowatt* (kW) – 10^3 W – and *megawatt* (MW) – 10^6 W. The latter are used in describing the outputs of power stations; but these are megawatts of *electrical* power, of course.

Power has to be applied for a period of time in order to convert it into energy, so a unit of energy called the *watt hour* (Wh) is used, and a power station produces *megawatt hours* (MWh) or *gigawatt hours* (GWh – 10^9 Wh) of electrical energy. To avoid further confusion, we speak of 'watts (thermal)' (Wth) or 'watts (electrical)' (We).

Joule's research may not seem particularly exciting today, but it was tremendously valuable because it proved experimentally for the first time the principle of the conservation of energy; internal molecular energy can be produced in a substance *either* by adding heat *or* by mechanical work. Joule's efforts showed that there really is a direct relationship between the two forms of energy. It also opened the way for a whole new branch of science – *thermodynamics*.

1. When a flame is applied to a metal object – e.g. a wheel – heat energy Q affects its molecules.
2. Friction on a revolving wheel can achieve a similar result, but no energy has been transferred due to a temperature difference so U (internal energy) is used.

3. But the work done in lifting the wheel is converted into gravitational p.e. without affecting the internal energy of its molecules.

Heat and motion

Thermodynamics is the study of the relationship between mechanical energy and heat, and today it can be applied equally to the calculations involved in the design of car engines, jet turbines, rocket motors, electric motors, electrical power generators, or any other mechanical device. The *First Law of Thermodynamics* is in effect another way of stating the Law of Conservation of Energy, for in an engine internal energy changes may take place both as a result of heat flow and as a result of work done by or on its material, but no energy is actually created or destroyed – merely changed. To avoid confusion between the various forms, engineers use special symbols: U for internal energy, Q for heat, and W for work. Using these, the First Law may be written as: $U = Q + W$.

Just as it is important to distinguish between weight and mass, so it is with heat (Q), which is used only to describe the transfer of energy from one body to another due to a temperature difference between them, and internal energy (U), which may *appear* as 'heat energy' when transformed into work. The *Second Law of Thermodynamics*, stated simply, says that 'heat will not move, unaided, from a cold object into a warmer one'. This seems obvious now, but was an important step. High-temperature energy can be put to greater use than low-temperature, because of this one-way flow.

In the 17th century the English scientist Robert Boyle and other researchers such as Robert Hooke regarded heat as being a form of motion – 'a very brisk and vehement agitation', as Hooke put it, which is quite close to the modern view of internal energy. But they had not yet linked this with the idea of work, and by the end of the century an entirely different theory had taken over.

Until about 1725 it was not even possible to measure temperature – making a scientist's job very difficult. Daniel Fahrenheit, following on the heels of Galileo and the members of an academy in Florence, devised a *thermometer* with a scale of degrees in which 32° was the temperature at which water froze and 212° that at which it boiled. In 1742 the Swedish scientist Anders Celsius proposed using 0° as the freezing point and 100° as boiling, and the *Celsius* scale – often known as *Centigrade* – (°C) has now been adopted into the SI system, replacing the *Fahrenheit* (°F) scale used until recently by countries such as Britain and the USA.* However, cold being merely a lack of heat, there must be a point at which all heat has gone; 'absolute' temperatures are measured in *Kelvin* (K) after Lord Kelvin, the English physicist. Degrees Celsius can be converted into Kelvin by adding 273, because $-273°C$ is considered as 'absolute zero'; so $0°C = 273 K$.

The science of heat measurement is called *calorimetry*. The term 'calorie', as a quantity of heat, of course came from the same source, and the story of that source makes interesting reading as an example of how scientists can sometimes take a wrong road – and travel up it for 150 years.

---

* To convert °F into °C: deduct 32, multiply result by 5, divide product by 9. To convert °C into °F: multiply by 9, divide product by 5, add 32 to quotient.

Under the old system, in calories per gram per degree C, the specific heat of water was 1.00. In the SI system – joules per kilogram per degree C (J/kg°C) – it is 4,200. Some other common substances are shown here for comparison.

Water	4,200	Iron	460
Sea water	3,900	Copper	400
Ice	2,100	Zinc	380
Aluminium	900	Mercury	140
Glass	670	Lead	130

Caloric

Next to the realisation that cold is only an absence of heat, and that heat flows from a warm object to a cooler one, the most important discovery in this field was that heat and temperature are *not the same*. This fact was deduced by a Scot, Joseph Black, who was working with Fahrenheit's new thermometer. Black found that various substances required different amounts of heat to raise their temperature by 1°F; he called this their 'capacity for heat'. A given mass of water, for instance, warms up much more slowly than the same mass of metal placed on the same flame. This capacity is now known as *specific heat*; water was given a specific heat of 1, and all materials are now given a value (almost all are lower than water, which means that water is slower to absorb or give out heat than nearly everything else – a useful property in many ways).

It becomes clear from what was said above that heat is something that can be 'put into' a substance, while temperature is a measure of its 'hotness'. Next, Black found that instead of becoming hotter and hotter, as had previously been assumed, as more and more heat was put into it, water reached a temperature of 212°F – boiling point – and *stayed* at that temperature until it was all converted into steam.

Likewise, he found that ice remained at 32°F until it had all become water, even though heat was being put in to melt it. He concluded that this was because, in the case of the boiling water, it was using all the heat to change from liquid into gas, and in the case of ice that the ice stayed solid until enough heat was put in to turn it into a liquid. This 'held in' heat he assumed to be somehow absorbed into the substance of the steam or ice-water, and he called it *latent heat*.

It is not surprising, then, that many scientists, including Black himself, discarded the 'motion' theory of heat, and instead thought of heat as a weightless fluid, which they called *caloric*. Particles of caloric were attracted by the particles of other substances, so they became hot. However, the particles of caloric repelled each other; this explained why materials *expand* when heated, as had been found by two Frenchmen, Jacques Charles and Joseph Gay-Lussac (independently) around

1800. This applies to solids, liquids and gases, and also explained why a gas, such as air, rises when it is warmed – it expands and becomes less dense. A car engine works because the gases heated by being ignited by the spark plug expand, pushing the piston down.

To return to caloric, it was necessary to assume that it combined chemically with particles of other substances, producing a new state of matter (e.g. water was ice combined with caloric, steam was water combined with caloric), in order to explain latent heat. It had to be weightless because, as the opponents of the theory were quick to point out, if caloric were *added*, a substance should weigh more when hot than when cold – and it doesn't.

Motion carried

Caloric's main opponent, around 1797, was an American, Benjamin Thompson, who was in the service of the Elector of Bavaria and was later given the title of Count Rumford. While in charge of the arsenal at Munich, he noticed the large amount of heat produced by the boring of a brass cannon. The caloric theorists would say that caloric was being squeezed out of the metal by the drill, presumably because the metal chips had less capacity for holding heat than the solid metal. But on measuring the specific heat capacities of both chips and solid metal, he found them to be the same. Perhaps the caloric came from the air? Rumford excluded the air by encasing the cannon in a wooden box filled with water and had a blunt drill turned by horses. As friction heated the brass the water warmed up, until after $2\frac{3}{4}$ hours it actually boiled. Since the water had *gained* caloric, this could not have come from the water. . . .

It was the beginning of the end for the calorists, and Rumford joined Boyle, Hooke, and later 'motion' proponents such as Davy, Mayer and Joule in having his theories accepted.

Humphrey Davy is said to have rubbed two lumps of ice together on a very cold day, finding that the friction melted them. Since the air temperature was freezing, caloric could not have come from that, and since ice has less capacity for heat than water the ice could not have supplied caloric for melting. The only conclusion was that heat was produced by the friction of motion.

The German physicist Robert Mayer, almost 50 years later, became interested in the heat produced when gases are *compressed*. If you have ever pumped up a bicycle tyre you will know how hot the pump becomes. At the time you may have put this down to friction, but an oiled plunger can do little work against a smooth barrel. The increase in internal energy actually comes from the work done in com-

Energy and the future

pressing the air inside. If Mayer had a bicycle it certainly did not have pneumatic tyres, but in 1842 he calculated the work done when a gas is compressed, assuming that the whole of this work was converted into internal energy, thus raising the temperature of the gas. This works in reverse too, for if compressed air or gas is allowed to expand it can perform work, the energy coming from the internal energy of the gas itself; which therefore cools. This principle is of great value in many of today's energy-producing (and -using) devices.

Robert Boyle discovered that by halving the volume of a gas (2) its pressure is doubled. Since its molecules now have half as much room they push twice as hard against the walls of their vessel, causing an increase in temperature. When the gas is allowed to expand (3), it cools. (Conversely, heating a gas increases its pressure.)

A matter of elements

We have looked in some detail at internal energy, which, as we saw on page 16, is caused by moving molecules. But what are molecules made of, and how do we know they are there? The term comes from the Latin word 'moles', meaning mass. Many of the ancient Greek thinkers believed that all substances were made up of tiny particles; around 400 BC Democritus called these microscopic balls 'atomoi', from which the modern word *atom* comes. The Greeks had the right idea, but it was not until the 19th century that it could be developed into a real scientific theory or proved experimentally.

Everything in the universe – stars, planets, Earth, your own body, this book – is composed of atoms. But atoms do not like to be alone; indeed, single atoms are very rare. They gang together, either with atoms of the same element or of some other element (an *element* being a substance which cannot be split into other substances). In either case they form molecules; but when the atoms of two or more different elements combine, they make molecules of a new *compound*.

An English schoolmaster, John Dalton, supported all this experimentally, and stated it in 1808 in his famous *Atomic Theory* (though the idea of elements had been suggested by Boyle over a century earlier). Dalton also showed that if each element could be composed of a *particular* atom, the idea that elements could

23

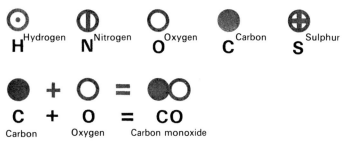
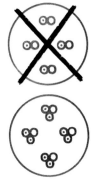

Dalton's symbols for some elements, and those of Berzelius as used today. It should be remembered that H is not merely a shorthand sign for hydrogen: it stands for 1 weight unit of hydrogen, while N stands for 14 of nitrogen, O for 16 of oxygen, and so on.

(Right) Dalton at first assumed that an atom of hydrogen and an atom of oxygen linked to make water. It was Henry Cavendish who who first showed experimentally that water consists of *two* volumes of hydrogen and one of oxygen.

combine to make compounds could be explained more easily.

Just as our knowledge of the true nature of heat was held up in the 18th century by the caloric theory, so chemistry was hindered by the *phlogiston theory* of a German, Georg Ernst Stahl, which was supported in England by Joseph Priestley. According to this, phlogiston was a mysterious element which made some substances combustible. When something burned, it was said to give up its phlogiston; this is why it crumbled into ash.

It was a Frenchman, Antoine Lavoisier, who, by accurately weighing substances before and after heating to ash or 'calx', showed that some actually *gained* in weight – which would hardly be expected if phlogiston had been given off. What actually happened was that the substance had combined with something in the air to make a new compound. Lavoisier coined the word *oxygen* for that something. He then experimented with other chemical reactions and found that in each case the weight of the resulting compound was exactly equal to the combined weight of the original chemicals. (This includes oxygen from the air, which appeared to add to the weight until the experiment was later done in a sealed container.)

By comparing the formulae of all the many compounds he knew, Dalton found that the weights which combined one with another seemed to have a definite relationship, which could be expressed in terms of the unit of hydrogen, the lightest; so he gave the atom of each known element an *atomic weight* (mass). That of hydrogen was 1, oxygen 7, and so on for the 30 or 40 elements then known.

He also gave each element a symbol, some of which are shown here, so could indicate compounds by putting these symbols together in various combinations. Today we use symbols introduced by a Swede who lived in Dalton's time, Jöns Berzelius, based mainly on the initial letters in the Latin name for each element: iron is 'Fe' (ferrum), lead 'Pb' (plumbum), tin 'Sn' (stannum), and so forth. Dalton hated these, and clung to his own.

Mixed reactions

John Dalton and other scientists found that whatever the proportions in which hydrogen and oxygen were *mixed*, they always *combined* in the ratio of 1 part (by weight) of hydrogen to 8 of oxygen (Dalton made it 7, which is why he gave this as its atomic weight) to produce water. The same applies to any other reaction – a fixed proportion of any given substance always combines with the same proportion of another substance to form a particular compound, though the proportions can vary greatly from one substance to another. The same elements may, however, combine in two or more differing proportions to form very different compounds;

In 1811 an Italian physicist, Amedeo Avogadro, discovered that equal volumes of all gases at the same temperature and pressure contain an equal number of molecules. He also proposed that gases – unlike liquids or solids – exist in 'diatomic' molecules (i.e. of two atoms each). Each oxygen molecule (of billions) in the top vessel weighs 16 times as much as a hydrogen one below, but each vessel contains the same number of molecules.

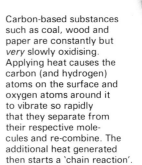

Carbon-based substances such as coal, wood and paper are constantly but *very* slowly oxidising. Applying heat causes the carbon (and hydrogen) atoms on the surface and oxygen atoms around it to vibrate so rapidly that they separate from their respective molecules and re-combine. The additional heat generated then starts a 'chain reaction'.

but the grouping of atoms is always the same in any specific compound. For instance, 3 parts by weight of carbon combine with 8 of oxygen to form relatively harmless carbon dioxide, but 3 of carbon can also take up only 4 of oxygen, making poisonous and inflammable carbon *mon*oxide.

Dalton went wrong with some of his atomic weights due to the fact that he did not at first realise that the atoms of some gases go about in linked pairs, which we also call molecules (so we normally write the symbol for oxygen as 'O_2', not just as 'O'). Once this was realised, scientists could express the formula '2 molecules of hydrogen plus 1 molecule of oxygen make 2 molecules of water' as: $2H_2 + O_2 = 2H_2O$. Since the oxygen combines with *two* molecules of hydrogen, not one, its atomic weight can now be seen to be 16, not 8. This ability to combine is known as *valence*.

There is not room here to go further into chemical reactions; any school chemistry book will cover the subject fully. The point for the purpose of this book is that we have another form of energy – *chemical* energy. Almost every chemical reaction produces energy, usually as heat; those that do are called *exothermic*. Others, needing heat put into them, are *endothermic*. Some need just a little energy, if only from a match, to start them. It is the chemical energy of wood, coal, oil and gas which keeps us warm and propels our engines, and the chemical energy of food which enables our muscles to do useful work, by combining with oxygen in the process of *oxidation*.

We still have not seen what an atom, or even a molecule, looks like. That molecules *are* in constant motion was proved by accident by a botanist, Robert Brown, in 1827 when he was examining very small pollen particles suspended in water, through a microscope. He was surprised to see that they were moving around rapidly, for no apparent reason. We now know that it was due to the impact of moving water molecules, and the phenomenon (which can also be seen in gases by using smoke particles) is known as *Brownian Motion*.

Special 'electron microscopes' which can magnify up to 5 million times can now show the patterns of atoms which make up a substance. Most solids show a regular pattern called a *lattice* which reveals their crystalline structure; their molecules alternately attract and repel one another. In a liquid (which can also result from applying heat to a solid until its lattice breaks up) the molecules vibrate to and fro in the same way, but they are also free to move amongst one another, exchanging partners. This is why a liquid takes up the shape of the container into which it is poured. In a gas, the 'third state of matter', the molecules are much farther apart than in a solid or liquid and move around rapidly, colliding with each other and bouncing off the walls of their vessel at perhaps 15 billion* collisions a second. Small wonder that they expand to fill their container completely!

---

* In this book the US convention: a billion = a thousand million; a trillion = a million million has been found the most convenient.

(Below) Bohr's model of the atom. The nucleus is shown here as a single ball.

(Right) Water is the only substance familiar to man in all three states. In solid ice the H_2O molecules are locked rigidly into hollow rings (causing ice's low density). As water, the molecules are free to slip closely amongst each other, allowing the liquid to flow. Steam acts as any gas. Holding gas and liquid molecules together are *Van der Waals forces*, named after the Dutchman who postulated them in 1873. These are due mainly to electrons temporarily collecting on one side of an atom, causing a negative charge on that side and positive on the other, thus affecting and attracting neighbouring atoms.

Ice

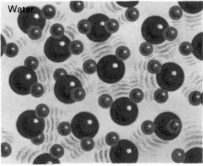

Water

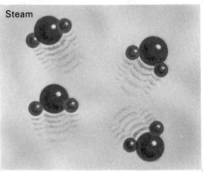

Steam

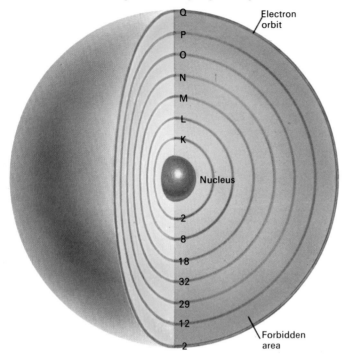

(Below, left) A JEM-100CX Transmission Electron Microscope and *(right)* a photograph taken at a magnification of × $2\frac{1}{2}$ million, showing the lattice pattern of atoms in graphitised carbon black. *Photographs by permission of JEOL (UK) Ltd.*

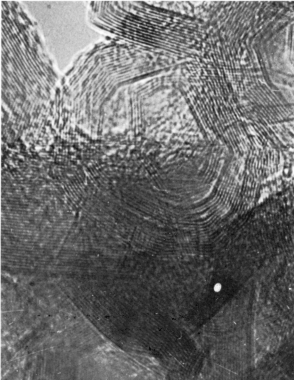

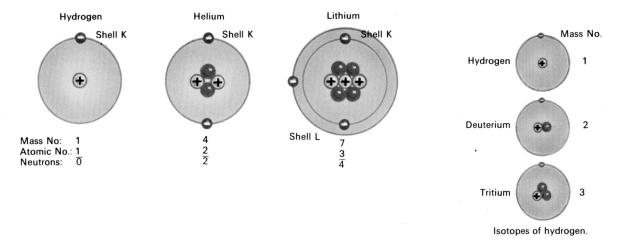

Isotopes of hydrogen.

Building an atom

As yet we can see atoms only as patterns of dots on photographs taken by special microscopes. Any models or drawings we make can only be greatly simplified, but they can help us to understand the behaviour of most matter and energy.

One of the earliest, and still most useful pictures came from a young Danish physicist, Niels Bohr. In 1913 he was studying under Lord Ernest Rutherford, who had already proposed a model of the atom. Bohr took this further, and saw the atom as a miniature solar system with a *nucleus* as the Sun and *electrons* whirling around it in circular orbits, like planets. This was modified by a German, Arnold Sommerfeld, who thought the orbits should be elliptical. However, it soon became apparent that the orbits could not all be in the same plane, as are the planets, but must form at least seven spherical 'shells' inside each other, like the skins of an onion; the spaces between the skins were 'forbidden areas'. And only a set number of electrons were 'allowed' in each shell, as shown in the diagram. Chemical reactions such as we saw on page 25 depend almost entirely on the number of electrons in the *outer* shell.

Vital in holding together an atom are its electrical charges. The nucleus, far from being a solid ball (though most of the atom's mass is concentrated into this), is composed of from one to 100 or more *protons*, which have a *positive* charge just as large as the charge on the electrons; but the electron's charge is *negative*. Since the number of protons is the same as the number of electrons, the atom as a whole is normally *neutral* electrically. However: hydrogen contains one proton. Helium contains two, yet is about *four* times heavier than hydrogen, not twice. Carbon contains six, yet is *twelve* times as heavy as a hydrogen atom.... A proton is nearly 2,000 times as heavy as an electron, so the weight difference can hardly come from the electrons – yet something seems to be missing.

Another of Lord Rutherford's co-workers, James Chadwick, discovered the missing particle in 1932. He called it the *neutron*, because it was neutral and so did not affect the electrical balance of the atom. Hydrogen would still consist of one proton with one electron going round it; helium would have two protons and two neutrons in its nucleus, with two electrons in orbit; the carbon nucleus would have six protons and six neutrons, six electrons balancing the positive charge of its protons. The nuclei of some elements can contain different numbers of neutrons, as shown, and are then called *isotopes*.

Some new terms had to be invented to describe the increasingly complex nature of the atom. The *atomic number* is the number of protons in the nucleus; so hydrogen is 1, helium 2, and so on. The *mass number* is the number of particles – protons *and* neutrons – which make up the nucleus. Again hydrogen is 1, but helium is 4, etc. The atomic number can also equal the number of electrons, and we have only to subtract the atomic number from the mass number to find the number of neutrons in the nucleus.

Dalton and Mendeléev (see next page) tried to keep hydrogen as the 'standard atomic weight' at 1 also, while another school of thought pressed for oxygen to be used at 16, which caused some confusion for up to 100 years; but in 1961 carbon was accepted as the international standard at 12, on which scale (for accurate measurements) hydrogen weighs 1.008 and oxygen 15.999. *Molecular weight* is simply the sum of the weights of all the atoms of which a particular molecule is composed.

The Periodic Table

In 1869 the Russian chemist Dmitri Mendeléev drew up his *Periodic Table*, with the aid of which, for the first time, scientists were able to find the atomic composition of any chemical and deduce the formulae of new compounds by calculation.

What Mendeléev did was to take 63 cards, on which he wrote the name, properties and atomic weight of all the elements known to him. He pinned these to the wall of his laboratory, and found that, with only a few exceptions, they could be grouped into just seven vertical columns, each containing elements with similar properties – light metals, heavy metals, non-metals with various characteristics, and so on. The hydrogen card did not fit into any of these, so he put it on its own.

As you see from the section of his chart here, he found to his surprise that when he rearranged the columns side by side so that their elements' increasing atomic weights could be read *across*, elements with similar properties appeared on every seventh card, so beneath one another vertically. You will notice that there is a gap under aluminium. Mendeléev knew that titanium, the next *known* element, did not belong there, because its properties placed it in column 4, under carbon and silicon, so he left the space in column 3 blank; then later *predicted* an element – which he named 'eka-boron' – to fill it. An element of the correct atomic weight and properties, now called scandium, was indeed discovered just ten years later; similarly with other gaps. The modern Periodic Table has 18 columns rather than seven, and there are still gaps towards the end. One of the most recently-discovered man-made elements (more about them later), Lawrencium, has 103 electrons in seven shells – i.e. its atomic number is 103.

The 'period' at which similar elements appear (which gave the Table its name) is obviously directly connected with their atomic number, but even more with the number of electron shells used by the atoms in a particular row. These shells build up their 'quotas' of electrons as shown. The innermost, shell 'K' on Bohr's model, could hold only two electrons – in hydrogen it has only one, in helium two; it is then 'full', so lithium has two electrons in shell K and one in L. Neon has a full quota of two in K and eight in L, making ten in all, so sodium starts on a third shell, M, with its eleventh electron. After the fourth row some atoms tend to fill outer shells before some of the inner ones are full, and complicate the picture, but on the whole the Periodic Table makes a logical pattern of electrons and their shells. And once all the gaps up to number 92 – uranium – had been filled, it seemed certain that no more *natural* elements would be discovered. Until, that is, in 1976 a team of physicists in America discovered not one but three new 'superheavy' elements, with atomic numbers of 116, 124 and 126, in a piece of Madagascan mica.

Somehow, the 'balance' of the atom as a whole seems to be determined by the maximum number of electrons in each shell. It does not seem to like an unfilled shell – especially an outer one – so if an atom sees a chance to link up with or 'steal' another electron which will complete its quota, it leaps at it. For instance, sodium, as we saw, has only one electron in its outer (third) orbit, i.e. one *more* than it needs for two complete rings. Chlorine has seven, and takes the odd one from sodium, so that both have complete rings, forming sodium chloride – common salt – and both are happy!

A section of Mendeléev's Periodic Table.

Group 1	Group 2	Group 3	Group 4	Group 5	Group 6	Group 7
Lithium	Beryllium	Boron	Carbon	Nitrogen	Oxygen	Fluorine
Sodium	Magnesium	Aluminium	Silicon	Phosphorus	Sulphur	Chlorine
Potassium	Calcium	?	Titanium	Vanadium	Chromium	Manganese

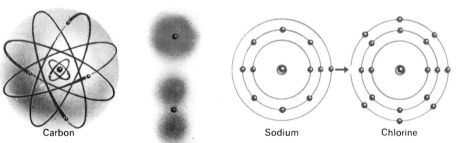

Carbon　　　　　　　　　　　　　Sodium　　　　Chlorine

The diagrams on page 27 suggest that several electrons occupy the same orbit. The carbon atom depicted three-dimensionally *(left)* is more accurate, though still strictly diagrammatic. Scientists actually think of an electron's path as a 'volume of probability' in which it may be at any instant of time *(centre)*. The region of highest probability is known as its *orbital*. At room temperature a hydrogen atom's orbital is spherical. At higher energy levels it becomes dumb-bell shaped. *(Right)* The single outer electron of a sodium atom fills the empty space in shell M of chlorine.

In a water molecule – a covalent compound – each hydrogen atom gives one electron while each oxygen atom gives two to complete the bond. Shown here 'in perspective', the hydrogen nuclei are actually held at an angle of 105° to the oxygen nucleus, keeping one side of the water molecule negative and the other positive, thus attracting other water molecules even more strongly than by Van der Waals forces, and giving water its unique properties, including 'wetting' and high boiling and freezing points.

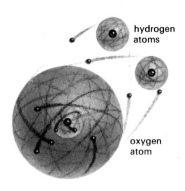

hydrogen atoms

oxygen atom

Attraction of opposites

The nucleus of an atom is (relatively) large and positively charged, the electron small and negative. 'Like' charges repel each other, opposites attract, just as with the poles of a magnet. But the electrons are whizzing around at some 2,500 kps, so just as planets do not fall into the Sun, electrons are not pulled into the nucleus – each stays at its own fixed distance. Their orbits must be very complicated, though, especially those with 90 or more electrons in seven shells. Since all the electrons are negative and repel each other, they try to keep as far apart as possible in their orbits at all times, so the balance of the atom is constantly changing.

When the chlorine atom completed its outer ring by collecting an extra electron from sodium, it acquired a negative charge (you can see why: the ratio of electrons to protons has changed). The sodium, having lost an electron, became positively charged. Negative attracts positive, so the atoms are held together; this is known as *ionic bonding*; an atom or molecule which has lost electrons is a *positive ion*, one that gains is a *negative ion*. Where *each* atom has several electrons in its outer shell but needs one or more to complete it, they may *share* each others; this is *covalent bonding*, and is usually stronger than ionic bonding. Atoms whose outer shells are complete do not react with other atoms in this way and are called *inert*. The most reactive atoms are those with only one electron in their outer shell.

As we saw earlier, the chemical energy which is released during a reaction is usually given off as heat. It is not possible to measure directly how much energy is bound up in the electrical forces which hold a molecule together, but of course we can measure the heat given off when a reaction occurs. A device for doing so, first made in the 18th century, is called a *calorimeter*. It consists simply of an insulated container full of water, with a small vessel inside in which the reaction takes place – often very rapidly. A thermometer measures the rise in temperature of the water in the outer container. Joule's paddle experiment was also carried out in a calorimeter, and even food is burned and the heat released measured in this way to find its Calorie content. Scientifically, the energy is today measured in joules (or kilojoules, etc.).

The formation of an everyday substance such as water can be quite startling,

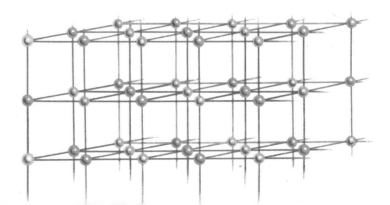

Sodium (yellow) and chlorine (green) ions form a cube-shaped crystalline structure in an ionic bond as sodium chloride. When salt dissolves, freely moving water molecules break up the lattice; it re-forms when the water is evaporated away.

for when a mixture of the gases hydrogen and oxygen is ignited they combine with a loud 'bang!', producing tiny droplets of H_2O – and releasing a great deal of heat energy. The second and third stages of the Apollo launch rocket used liquid hydrogen (the gas liquefies only when cooled to $-253°C$) and liquid oxygen (at $-183°C$) as propellants, which is a measure of their efficiency as energy-producers. From extreme cold, the temperature in the combustion chamber reached over 3,000°C; it requires a similar temperature to break up water (or virtually any other stable compound) into its component elements again. Yet water can quite easily be decomposed into hydrogen and oxygen by *electrolysis*; in reverse, these gases can produce electrical energy plus water in the special 'fuel cell' also developed for Apollo.

Making the sparks fly

Water was first split into its elements by electrical means around the beginning of the 19th century by two London professors, Sir Anthony Carlisle and his friend William Nicholson. Before they could do so, a chain of events had to take place which led to the discovery of electricity.

For thousands of years the 'magical' properties of natural magnets such as the 'lodestone' had been known, and had puzzled thinkers. I have already mentioned the opposite poles (known as 'north and south' because a suspended bar magnet always lines up pointing in these directions – hence its use in a compass) which attract each other, whereas like poles repel.

The Greeks knew of another – apparently similar – phenomenon which occurred when amber was rubbed. It attracted small particles such as dust and hair; other materials do the same; from sealing wax to some modern plastics out of which combs, pens and gramophone records are made (as 'hi-fi' addicts know only too well). Everyone also knows of the crackle (and even spark, in a darkened room) when nylon clothes are removed on a dry day, or when hair is combed. These are all signs of *static electricity*; the Greek for amber is 'elektron'.

In 1672 a German, Otto von Guericke, found that feathers were first attracted to, then repelled by a ball of sulphur charged by rubbing – and that a thread attached to the ball also transmitted this force. In France, 150 years later, Charles Du Fay realised that there were two kinds of static electricity, each type attracting the other but repelling its own kind. He called these 'vitreous' and 'resinous' after the sort of material producing them, but in America Benjamin Franklin introduced the terms *positive* and *negative* which we use today.

Franklin, who risked death by electrocution when he flew a kite in a thunder-

A Wimshurst machine (designed by British engineer James Wimshurst in 1881) for producing static electricity by the friction of two glass discs rotating in opposite directions. Electrons are 'stolen' from the tinfoil lining of one Leyden jar and transferred to the other, giving one a highly positive and the other a negative charge. When the two upper knobs are brought close together a crackling spark discharge passes between them. (*Above*) On a giant scale, similar forces in nature cause lightning.

cloud and drew a spark to his finger from a key attached to the string, believed that electricity and magnetism were fluids – just as heat was thought to be. Anything which had more than its fair share of electrical fluid had a 'plus', so was positive; anything which had less was 'minus' or negative. Sir William Watson had the same idea in England, but Franklin got the credit.

The first use of a spark to ignite inflammable substances or explosives was in 1744 (the substance was French brandy). The same principle is used today in any car engine. In 1745 Sir William Watson perfected (though did not invent) the *Leyden jar*, in which electrical charges could be 'condensed' or accumulated. From these, electricity was conducted by a wire for up to two miles (3.2 km), and it was found that *earth* would act as a return path. Attempts by Sir William to measure the speed of the charge down the wire suggested – incorrectly – that it was instantaneous.

In 1780, at the University of Bologna, Luigi Galvani was dissecting a frog when an assistant in the same laboratory made a spark on a static electricity machine. As he did so, the frog's legs jerked just where Galvani's metal scalpel touched. Fascinated, Galvani performed similar experiments, one of which was to attach a dead frog's legs to a brass hook and hang them on an iron fence during a thunderstorm to see if they would twitch. They did, whenever they touched the fence – whether there was any lightning or not.

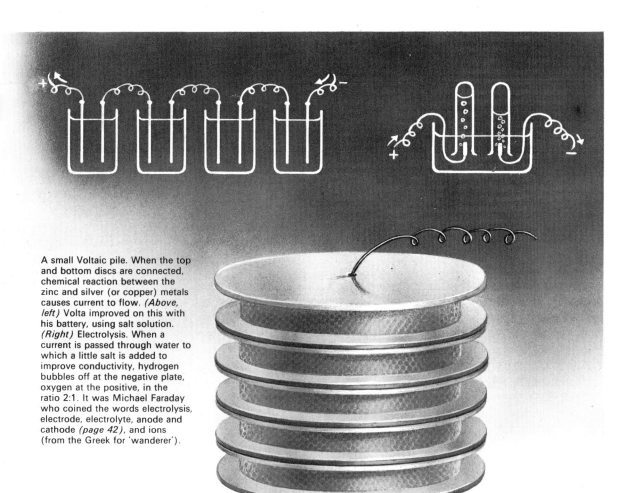

A small Voltaic pile. When the top and bottom discs are connected, chemical reaction between the zinc and silver (or copper) metals causes current to flow. *(Above, left)* Volta improved on this with his battery, using salt solution. *(Right)* Electrolysis. When a current is passed through water to which a little salt is added to improve conductivity, hydrogen bubbles off at the negative plate, oxygen at the positive, in the ratio 2:1. It was Michael Faraday who coined the words electrolysis, electrode, electrolyte, anode and cathode *(page 42)*, and ions (from the Greek for 'wanderer').

The current flows

Professor Galvani's conclusion was that the frog's legs contained 'animal electricity'. Another Italian professor, Alessandro Volta, followed this up in 1800 by showing that the frog's legs could be replaced by cloth or paper soaked in salt solution. He made the world's first *battery*, known as a 'Voltaic pile', by interleaving discs of cardboard soaked in brine with alternate discs of silver and zinc. Instead of the single discharge produced by static machines, electricity flowed continuously along a wire attached to the top and bottom metal plates, the first time a steady *current* had ever been made. Later experiments showed that other combinations of various acids or salts, such as zinc and copper in dilute sulphuric acid, could be used. Such a *simple cell* is easily made.

It was while Carlisle and Nicholson were experimenting with the new voltaic pile that they discovered electrolysis. Sir Anthony had placed a drop of salt water on the top disc – which was a silver half-crown – to improve the electrical contact with the wire he was about to connect, when he noticed that bubbles of gas rose from the wire when it was merely dipped in the water. It was an important discovery, and Sir Humphrey Davy later isolated the metals potassium and sodium by passing a current through molten compounds of the metals, which do not appear in nature because they are so reactive that they burst into flame on contact with water. One of the earliest industrial applications was *electroplating*, with which a base metal such as iron is coated with a more expensive one such as chromium or silver.

The word 'current' stems from the time when electricity was thought to be a fluid, but it is still useful to think of it as a flow. As with heat, electricity is connected with *motion*. We already know that an atom is composed of positively-charged protons and negatively-charged electrons in equal numbers, showing that electricity is a vital part of matter, not separate from it. When glass is rubbed with silk, some electrons loosen themselves from the glass and attach themselves to the silk; so the glass becomes positively charged and the silk negatively. When a stick of sealing wax is rubbed with flannel, electrons fly *from* the flannel *to* the sealing wax, which becomes negative while the flannel becomes positive.

Early in the 17th century Dr. William Gilbert, physician to Queen Elizabeth I, had listed a number of substances that became electrified by rubbing, which he called 'electrics'. He also listed 'non-electrics' which were not affected, most of which were metals. When current electricity was discovered 200 years later, current was found to flow through non-electrics but not through electrics. The former are now called *conductors* and the latter *insulators*. In an insulator the electrons are bound tightly in their atoms, while the electrons in a conductor are free to move from one atom to another, so allow a free flow of current within their substance. When an insulator such as glass is held in the hand and rubbed, the charge of electrons is formed on its *surface*. They cannot flow to earth through the hand (unless it is damp) because the material is an insulator; when a piece of metal is rubbed a charge *is* formed on its surface, but is immediately conducted through the body to earth which is always at a potential to accept it.

Just as water flows more freely in a wide pipe, so electrical current moves more easily through a thick wire – though material is important too. And as water flows downhill, and heat flows from a hot object to a colder one, so an electric current travels from a place where there are many electrons to one where there are few.

Names to remember

Alessandro Volta's name is remembered when we measure the 'pressure' of electric current through a wire (scientifically: the 'potential difference' between the ends of a wire; remember, the current tries to even up the number of electrons at each end) in *volts* (V). The higher the voltage of a battery, the greater the flow of current. As we saw, a thicker wire offers less resistance to the flow of electrons than a thin one. A German physics teacher, Georg Simon Ohm, proved this in 1826 and has been honoured by having his name given to the measure of electrical *resistance* of a material – the *ohm* (Ω). So: current = voltage ÷ resistance (in ohms).

The 'unit of quantity' of electricity is named after the French physicist Charles

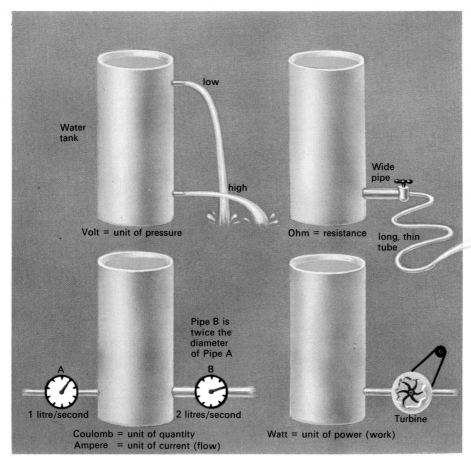

The terms used to describe electricity may be more easily understood by analogy with a flow of water. Even so, the movement of electrons (*not* the atoms themselves) in a circuit can be more accurately compared with firemen passing buckets in a 'chain' — except that an electrical circuit has to be completed. A battery or generator causes electrons to move in one direction — just as water in a pipe will not flow until the tap is turned on — by causing an *electromotive force (emf)* as volts.
Oddly, although by an early convention a direct current is said to flow from positive to negative, the electrons actually travel the opposite way!

Augustin Coulomb. The *coulomb* is 'the quantity of electricity which passes any point in a circuit in one second when a steady current of one ampere flows'; or, more simply, coulombs = amperes × seconds.

At this point it is as well to realise that although we speak of 'current' and 'flow', an electron does *not* start at one end of a wire and travel down it until it reaches the other end. The electrons are passed along from one atom to another, more like a game of 'pass the parcel'. In a wire they may travel as slowly as a few centimetres a second, but in gas they move at up to 100,000 kps – literally 'as quick as lightning'. The *ampere* (A), used above, is the unit of the current flow itself, named after another Frenchman, André Marie Ampère, who first likened electric current to a flow along a wire. So the ampere (often shortened to 'amp') is a rate of flow of one coulomb per second, which is similar to measuring the flow-rate of water in litres per second. Actually, 6,242,000 *trillion* electrons pass the measuring point in the wire for one ampere to register; but it is normally measured by the magnetic force the current exerts (in newtons).

We saw on page 19 that the power of any electrical appliance (bulb, fire, amplifier, etc.) is measured in electrical watts (We), after James Watt. We can therefore now say that watts = volts × amperes. If there is a potential difference of one volt between the ends of a wire, one joule of work is done per coulomb of electricity.

You can see that it is quite possible to have a very low voltage with a high amperage, if a lot of electrons flow at 'low pressure'; or a high voltage with low amperage, such as is generated by a static electricity machine or an induction coil. A toy 'shock coil' can produce several thousand volts, but does no harm to the person

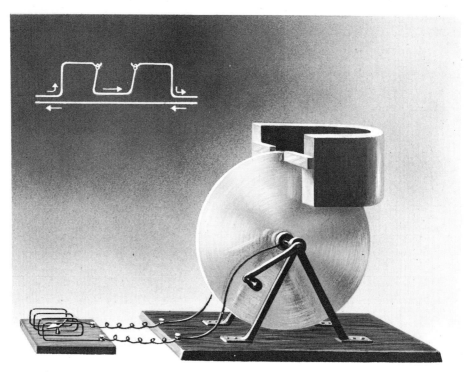

Faraday's copper disc 'generator'. It had two rubbing contacts and was rotated between the pole pieces of a permanent magnet. A galvanometer needle, as invented by Schweigger, registered current as long as the disc was turned. *(Above)* Ampère's experiment with attraction and repulsion in wires. The centre section is freely hinged.

holding the handles because the amperage is so low. 15,000 volts is necessary to make a spark jump a 25 mm gap, but lightning produces both millions of volts and thousands of amps, which makes it so dangerous. Incidentally, it was as a result of experiments such as that of Benjamin Franklin that most tall buildings now have a *lightning conductor* – a strip or rod of metal (usually copper) which conducts the charge from the earth, so discharging the thundercloud harmlessly.

Silver offers least resistance to current (which means that its electrons are the most easily 'loosened'), but as this is so expensive, copper is the next best. These are called 'good conductors', while insulating materials are 'poor conductors'. Sometimes a high resistance is needed to reduce the flow of current in the circuit of, say, a radio or TV set. *Resistors* made of special alloys such as constantan are used for this purpose.

Putting electricity to work

There is one other effect of resistance that has not yet been mentioned: heat. The electric bulb and fire glow and give out heat because of the high resistance of the fine metal-alloy wire of their filaments. The usefulness of this in modern life is obvious. The resistance of pure metals actually increases with temperature (and decreases almost to zero if the metal is chilled – but that is another story), though that of some materials, such as carbon – a non-metallic conductor – decreases. It was a carbon filament, in the form of a charred cotton thread inside a hollow glass ball from which the air had been removed, to prevent oxidation, which Thomas Edison used in his first successful light bulb in 1879.*

Elsewhere in this book we have seen that one of the most useful transformations is between heat energy and mechanical energy. Obviously it *would* be possible to use electrical resistance to heat water to steam, or a gas such as air, and use the

* Sir Joseph Swan actually demonstrated this in Britain some 8 months before Edison in America, who patented the process first.

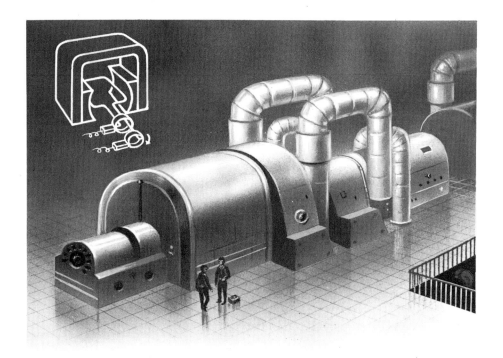

The simple device opposite led to today's massive modern generators at power stations, operated by steam turbines via the great pipes seen here. *(Above)* A simple AC dynamo. A rectangular coil of wire rotates between the poles of a *field magnet*; the ends of the coil are connected to two *slip rings* on the coil spindle. When the two carbon *brushes* are connected into a circuit and the coil is rotated, current flows. To produce direct current, the slip rings and brushes are replaced with a single split ring or *commutator* with a brush at each side. If instead these brushes are connected to a battery the coil will rotate and the dynamo becomes an electric motor.

expansion and compression of this to drive some form of engine. But this would be an inefficient way of doing it, to say the least! For whereas a steam engine converts only some 30% of the energy put into it into work, and the petrol engine 50%, electrical energy can be converted directly into mechanical energy with over 90% efficiency. To find out how, we have to return to our magnet.

In Copenhagen, in 1819, Professor Hans Christian Oersted found that a magnetic compass needle was influenced by being placed near a wire in which a current was flowing but only if wire and needle were parallel; he failed earlier when he placed them at right angles. The following year, in Paris, Ampère showed that two wires placed side-by-side with a current flowing in the same direction in each attracted one another, whereas if the current in one wire was reversed they were repelled. The connection between electricity and magnetism was apparent.

A further ten years passed before an even more significant event – the production of an electric current *from* a magnetic field – took place. Although Michael Faraday, in England, usually receives the credit for producing electricity by *electromagnetic induction*, in 1831, he was actually a year behind Joseph Henry in America, who failed to publish his results at once. It matters little; the invention has affected our whole way of life by making possible, as well as the electric motor, the electricity *generator* which produces such a vast amount of our energy today, and works in basically the same way whether it is powered by the kinetic energy of water or wind, by coal, oil or nuclear fuel.

On 17th October, 1831, Faraday coiled copper wire and string (as insulation) alternately round a paper tube, with strips of calico between each layer of wire. He connected the ends of the wire to a *galvanometer* (named after Galvani, though it was invented by a German, Johann Schweigger), which indicates a flow of current. When he pushed a bar magnet into the coil the galvanometer needle jerked, then went back to rest until the magnet was removed, when it kicked again.

Energy and the future

Faraday decided that current was induced in the coil due to the wire cutting through the magnet's lines of force. He performed many similar experiments, finally making a machine consisting of a 12-inch (330-mm) diameter copper disc mounted so that it could be turned by a handle, between the poles of a magnet. A current was generated, and conducted away, as long as the disc was turned. This simple machine led to the modern *dynamo* and motor.

Let there be light

Although we have looked at energy being released by various processes as heat, it will not have escaped you that there is often another by-product – *light*. Our caveman produced heat by friction which eventually caused chemical combustion and light from the flames of his fire, but friction *can* raise a substance to white heat without actual combustion taking place. When we burn paper, wood, a candle, coal or oil, the molecules reacting with oxygen from the air release light as well as heat. So do those in the wire heated by electrical resistance – without oxygen, in the vacuum of a light bulb. Another form of light, known as 'cold light', is that from a *discharge tube*: the tube contains gas, such as neon, at very low pressure and current at a high voltage is passed through it. This causes ions in the gas to move about rapidly and collide with gas molecules, emitting light of a colour which depends on the gas – with neon it is red.

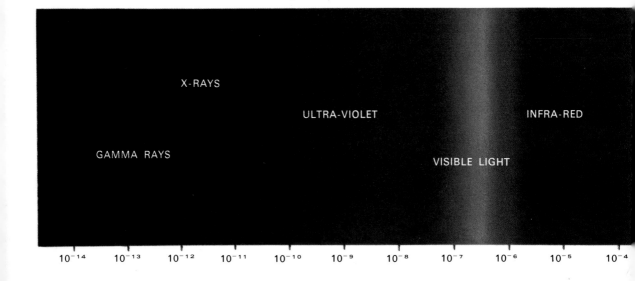

The electromagnetic spectrum, showing also the 'windows' by which light and certain radio waves reach Earth through the atmosphere.

So how does light fit into the energy scheme? I have explored the subject in some detail elsewhere*, but here, briefly, is the picture. From the Greeks, 500 years BC to Isaac Newton in the 17th century, light was regarded as a stream of tiny particles or 'corpuscles' which travelled in straight lines. The Greeks argued whether these were sent out by the eye or given off by the object looked at, but Newton knew that light itself is invisible and that an object can be seen only if it either *emits* or *reflects* light. The opponents to the corpuscular theory were Francesco Grimaldi in Italy and Christian Huygens in Holland, who believed that light was a form of *wave* in the 'ether' which was supposed to fill all space.

In England, Thomas Young performed experiments in 1801 which seemed to clinch the wave theory. He produced 'interference patterns' by recombining a split beam of light, which could be explained only by the wave hypothesis. Just about the same time, two new forms of light which could not normally be detected by the human eye had been discovered: *ultra-violet* (which causes sunburn) by a German, Johann Wilhelm Ritter, and *infra-red*, which can be felt as *radiant heat*, by the British astronomer Sir William Herschel. Newton had split white sunlight into a *spectrum* of seven colours (which we now know can be resolved into three basic colours – red, green and blue – unlike the primary pigments in paint etc., which are red, yellow and blue), so infra-red and ultra-violet could be fitted in at each end of this, but the great Scottish physicist James Clerk Maxwell went even further.

In 1873 Maxwell published a book in which magnetism, electricity, heat, light and radio waves, amongst other things, were linked. His *electromagnetic spectrum*

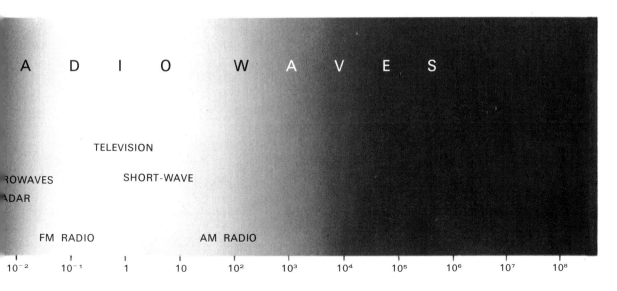

* *Light and Sight* by David A. Hardy (World's Work, 1977).

Energy and the future

shows how various forms of *radiation* are simply shorter or longer *wavelengths*, from gamma rays and X-rays at the 'short' end to very long radio waves at the other. A German, Heinrich Hertz, proved experimentally that radio waves existed, in 1888, when he made a spark jump across the terminals of apparatus across the room from the spark machine he was using. (This reminds us of Galvani's first experience with the frog, which, unlike his later 'galvanic' charges produced by two different metals, was due to *electrostatic induction* when negative electrons from the air within the spark gap induced movement in the electrons of the scalpel.)

It was in 1900 that another German, Philipp Lenard, made a discovery which seemed to show that light contained particles after all.

Making the quantum jump

Lenard knew of the discovery two years earlier by Wilhelm Hallwachs that electricity could be produced by shining a beam of ultra-violet light upon a sheet of zinc. Lenard realised that it does this by knocking electrons off the metal's surface. He found that although ultra-violet light produced the strongest effect, increasing the intensity of the beam did not speed up the electrons (known as *photoelectrons*) leaving the metal, though it increased the number. He also found that some other metals produced the effect with visible light, but only one – rubidium – with red light and none with infra-red. This *photoelectric* effect has many modern uses, but at the time it seemed that light was acting more as particles than waves.

Around the same time, also in Germany, Max Planck suggested that energy does not in fact flow steadily from hot bodies but is sent out in tiny 'packets' which he called *quanta*. (He had no evidence for this at the time, but it became a useful way to think of the energy of heat radiation; he was even able to calculate precisely the amount of heat in each quantum.) The mathematical genius Albert Einstein took this even further by saying that light too was made up of quanta, which were later called *photons* – the photons of each colour containing a different amount of energy. The photons in blue light (short wavelength) have more energy than those in longer red, whether it is bright *or* dim; and those in ultra-violet more still.

Nowadays, we *think* of some parts of the electromagnetic spectrum – from radio to visible light – as waves, and the very short end – gamma and X-rays – as particles, because of their behaviour; while visible light sometimes behaves as either. All travel at almost 300,000 kps. But remember that there are no actual particles – just little 'pulses' of energy in the electromagnetic wave. Proof that photons possess mass and momentum came in 1923 from an American, Arthur H. Compton. That comets' tails were somehow 'pushed' away from the Sun had been known in the

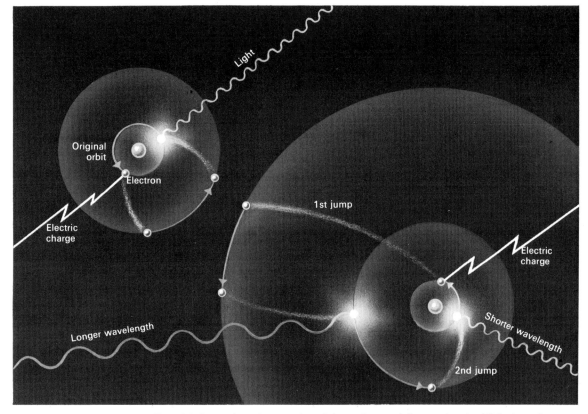

How excited atoms make light. Since it is further from the attraction of the nucleus, and the wavelength of light depends upon the energy involved, an electron which jumps more than one orbit will emit light of longer wavelength on its first step back than on the second.

17th century; this could now be explained by the photon-pressure in sunlight.

All this has not explained *how* light is produced. To find this out, let's return to Niels Bohr. When we left him on page 27 he had shown how electrons orbited the nucleus in definite shells instead of just a jumble. But what gave him this idea? He had become interested, too, in Planck's idea that radiation consisted of separate parcels rather than a continuous stream. Bohr knew that it is the electrons in any glowing object that give out the energy quanta of heat and light, and decided that he would try to do for the atom what Planck and Einstein had done for light; he would arrange its electrons in separate energy packets – the shells.

The energy an electron possesses as long as it spins in its own orbit is normally constant. This is known as its *ground state*; the electron does not give off energy or absorb it. But if additional energy is added – say, in the form of an electrical discharge, or as heat – it becomes *excited*. This is the scientific term, but it seems that it does literally jump for joy, for it leaps up to an outer shell. However it cannot stay there, because it doesn't 'belong', so it jumps back – and gives off the energy it had gained as a photon or quantum of light. Sometimes it may jump more than one orbit, in which case it comes back in steps, a shell at a time. The leap from the outer shell will give off a photon of lower energy than the next, to an inner one (because it is further from the nucleus), so gives off light of longer wavelength.

Now multiply this by the billions of atoms in the white-hot filament of a light bulb, and you see why it is so bright. The photo-electric effect is rather the reverse of the above, as Einstein showed: an electron of a particular metal requires a certain minimum amount of energy to loosen it, and *must* receive this in a single photon. Several 'smaller' photons will not do.

Energy and the future

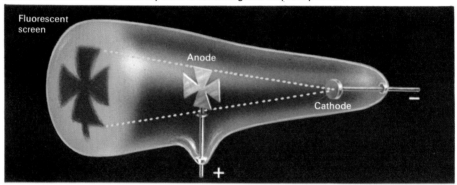

Plücker's Maltese cross cathode ray tube. It contains gas at very low pressure.

A new light on atoms

It was known in the 18th century that the crackling sparks from a static electricity machine could be changed into a quiet and beautiful glow if the *electrodes* at the end of each wire were encased in a glass tube from which the air had been pumped. We met this on page 38 as a discharge tube, as used today for advertising signs.

At Bonn University in 1859, Julius Plücker made a special tube with its positive electrode or *anode* in the shape of a Maltese cross, which cast a sharp shadow on to the green *fluorescence* of the glass at the end of the tube, showing that a beam of radiation was travelling in a straight line from the *cathode* or negative electrode. In England, Sir William Crookes and Sir J. J. Thomson constructed these *cathode ray tubes*, and concluded that what appeared to be a ray was actually a stream of high-speed particles. Over the next 40 years, among many experiments by these men and by Eugen Goldstein, Jean Perrin, Wilhelm Wien and others, the energy of the rays was made to turn a mica 'paddle-wheel' inside the tube; their path was made visible by a screen coated with zinc sulphide; they were bent by a bar magnet (which could not bend a light beam); and generally put through the hoop.

The German physicist Wilhelm Röntgen found in 1895 that covered photographic plates left near a working tube became fogged. He named the invisible radiation which came from the fluorescing glass wall of the end of the tube *X-rays*, and on placing a sheet of card coated with barium platinocyanide in the path of a beam found that it glowed brightly. He placed his hand between the tube and the card, and saw the shadow of his hand – with the flesh only faint but the bones clearly visible! We all know the value of X-rays today in hospitals and industry. These rays could not be bent with a magnet, and are not charged particles but, as we saw, very short electromagnetic waves. Another development of the cathode ray tube now forms the heart of every television set.

It was while working with a cathode ray tube with a hole in its cathode that Thomson noticed a stream of *positive* particles travelling in the opposite direction to the cathode rays and passing through the hole, making a bright spot on a screen at the end of the tube. Very careful measurements showed that these particles had different masses which depended upon the trace of gas left in the tube when most had been pumped out. These masses corresponded with the atomic weights of the gases. The positive stream consisted of atoms of the gas from which the (negative) electrons had been stripped. The cathode 'ray' was composed of those electrons. In separate tests, atoms of hydrogen, helium and oxygen had been pulled apart inside the tube, each into two sets of particles whose charges *balanced each other*. Thomson built up his first picture of the atom, as a sphere with equal positive and negative charges; and because the mass of the hydrogen (the lightest) without its electron was hardly any less than the mass of the whole atom, he deduced that the positive part was much the heavier.

You are probably ahead of me when I say that this experimental work led to the theory of the atom's structure proposed by Lord Rutherford and Niels Bohr. But Rutherford wanted to know a lot more about the atom. And once the secrets of the atom were probed, the way was open to the most powerful source of energy of all.

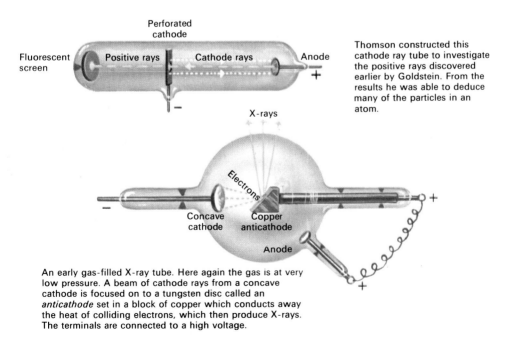

Thomson constructed this cathode ray tube to investigate the positive rays discovered earlier by Goldstein. From the results he was able to deduce many of the particles in an atom.

An early gas-filled X-ray tube. Here again the gas is at very low pressure. A beam of cathode rays from a concave cathode is focused on to a tungsten disc called an *anticathode* set in a block of copper which conducts away the heat of colliding electrons, which then produce X-rays. The terminals are connected to a high voltage.

Enter radioactivity

Up to now we have examined five apparently separate, yet now clearly interwoven forms of energy: *mechanical* energy, by which any machine works; *heat* and internal energy, which are often produced by or converted into the first; the *chemical* energy of fuels and food; *electrical* energy and magnetism; and *light* with its radiant energy cousins. All obey the Law of Conservation of Energy, linked by Einstein with the *Law of Conservation of Matter* (which states that the total amount of *matter* in any closed system – e.g. the universe – is fixed and cannot be increased or decreased) when he stated that 'the mass of a body is a measure of the quantity of energy contained in it' as part of his difficult Theory of Relativity – and actually simplified our view of the universe, when one thinks about it, as well as providing the Greeks' 'unifying principle'. However, the truth of this does not really become apparent until we look at the sixth and final form of energy – *nuclear energy* (often incorrectly called atomic energy).

We have examined atoms and molecules vibrating, joining up, losing or gaining electrons and positive or negative charges, and throwing off energy as heat, light or electricity. Certainly *all* this energy comes from the atom, in a sense – but it is chicken feed compared with that locked up in the atomic nucleus itself, which has not been affected by any of the methods we have looked at so far. To find out how this can be released we need to look at some elements near the end of the Periodic Table, one of which – uranium – I mentioned earlier in passing as the 'last' naturally-occurring element.

We saw that some elements, such as potassium and sodium, do not occur in nature because their electrons react so readily with water, or oxygen or some other element to form a compound. They are, however, 'natural' elements, and may be kept quite stable under the proper conditions. Other elements do not appear naturally because their atoms constantly give off bits of themselves or *decay*, which they do normally in two ways – *alpha decay* or *beta decay*.

In 1896 a French scientist, Henri Becquerel, was interested in the fluorescence of certain minerals, caused by cathode rays – just as were the men in the last two pages. One of these substances was a uranium salt, and he found that this fogged a wrapped photographic plate – even when the uranium had *not* been exposed to cathode rays, or even to sunlight, with which he had also been working. Whatever penetrating rays the substance gave out clearly did not require fluorescence. Becquerel had discovered *radioactivity*.

In Paris, two scientists who were also husband and wife, Pierre and Marie Curie, took an intense interest in Becquerel's discovery. They found that the rays ionised air molecules, and from this devised a way to measure their intensity. Another substance they tested which also emitted the ionising radiation was thorium, but an ore of uranium called *pitchblende* proved the most active. They were given a ton of pitchblende, a waste product of the Austrian Government's refineries, and from this painstakingly extracted a small amount of two *new* radioactive elements – *polonium* (after Marie's country of birth, Poland) and *radium*, more active still.

Becquerel, among others, found that two types of ray were given off by these elements, and yet another French scientist, P. Villard, discovered a third kind. Rutherford, then in Canada, named them *alpha* (α), *beta* (β) and *gamma* (γ), the first three letters in the Greek alphabet.

Rutherford, in separate experiments, subjected the rays emitted by a small piece of radium to a very strong magnetic field. The results were the same as in this experiment using two oppositely charged metal plates. By using photographic plates which act as radiation detectors, the amount of deflection (and thus electrical charge) could be seen. Gamma rays, being neutral, ignored the charged plates.

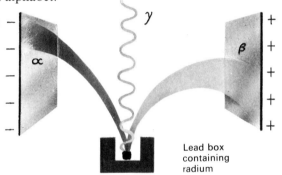

Lead box containing radium

Splitting atoms

The apparatus first used by Rutherford to study these rays was very simple, as shown above. Alpha rays were bent towards the negative plate, gamma went straight on, while beta were deflected even more strongly, to the positive plate. With more complicated apparatus, it was established that:

Alpha rays are composed of two protons and two neutrons, which come from the nucleus. This is the same as a nucleus of helium, though positively charged because it has no electrons. Having lost an alpha ray, a radioactive substance turns into a different element, two places down on the Periodic Table. The rays travel several centimetres in air, but most can be stopped by a sheet of aluminium foil or even paper.

Beta rays are high-energy electrons travelling at up to the speed of light, similar in fact to cathode rays. But these electrons or *beta particles* are ejected from the *nucleus* in some way not yet fully understood (producing at the same time an *antineutrino*, which need not concern us here), while – having lost a negative particle – a neutron turns into a proton and the nucleus acquires a positive charge. This converts it into the next element 'up' in the Periodic Table. Some beta rays can penetrate several millimetres of aluminium.

Gamma rays really are rays, or electromagnetic radiation, similar to X-rays but produced by energy changes in the nucleus, while X-rays originate from changes in the electron shells. They can penetrate several centimetres of lead.

Energy and the future

The atomic bomb dropped on Nagasaki on 9th August, 1945, expanded from a 550-metre fireball into a 6.5-kilometre high cloud. Devastating though this was, later nuclear weapons (hydrogen bombs) produced clouds 25 times larger. *(Above)* The sequence of events in a nuclear explosion in air. Fireball at millions of degrees gives out radiation (1), expands, causing a tremendous shock wave (2), then rises (3), sucking up dust and rubble and forming (4) the familiar mushroom-shaped cloud.

The Cloud Chamber devised by C. T. R. Wilson (originally to show how clouds condense around small nuclei) uses a piston to suddenly expand air saturated with water vapour. If a radioactive source is placed near a hole in the side its particles collide with air molecules and knock off electrons, leaving a trail of positive and negative ions whose paths are revealed when vapour condenses on them. The inside of the cylinder is matt black, but a bright light from the side enables tracks to be seen clearly or photographed, as shown at right.

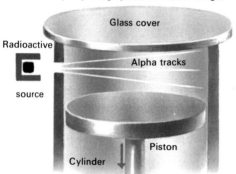

When an atom gives off alpha or beta particles and changes into another element, this element too is radioactive and disintegrates into yet another element, and so on. This is known as a *transformation series*. For instance, the isotope of uranium which occurs in nature, U-238 (the number indicates the atomic weight, remember), passes through 15 steps which include thorium 230, radium-226, radon-222, three isotopes of polonium, two of bismuth and three of lead, finishing up as a stable isotope of lead – lead-206.

The speed of the change is measured by *half-life*, which is the time taken for half of the atoms in any given sample to decay. If the half-life of element 'a' is one year and we start with one gramme, then 0.5g will have changed into element 'b' after a year. A year later another half, – 0.25g – will have decayed, and so on. There will always be a smaller and smaller amount left, until the final atom has changed. The fact that the half-lives of natural radioactive isotopes are known so accurately makes it possible to 'date' rocks etc. in which they occur. U-238 has a half-life of 4,510 million years, while many artificial isotopes last only a fraction of a second; others may be almost anywhere in between. It is not known *why* an atom chooses a particular instant to disintegrate, but it happens whether its element is pure or combined into a compound, and regardless of temperature.

In 1919 Rutherford and his team tried firing alpha particles through nitrogen gas at a zinc sulphide screen – smashing the nitrogen atoms' nuclei and producing hydrogen nuclei or protons *plus* oxygen nuclei. His feat as the first person to 'split the atom' was widely acclaimed.* He had turned the dream of every alchemist of the Middle Ages into reality, by *transmuting* one element into another – though their interest was in turning base metals such as lead into gold.

With the aid of photographs taken with the *Wilson Cloud Chamber* shown here, the evidence for the theories of Rutherford and those before him was almost complete.

* Though since the oxygen nucleus contains eight protons to nitrogen's seven he could equally have been said to have 'built' atoms.

War on the nucleus

Gradually, a whole new armoury of devices to bombard the nucleus with particles was built up. The main principle was to accelerate particles by means of very high voltages, as in the Cockcroft–Walton *proton accelerator* of 1932, in which the protons from a hydrogen discharge tube hit a plate of lithium at some 8,000 kps, and the fragments were sent into a cloud chamber. There was a slight loss in the total mass of the fragments, compared with the original nucleus, and this was shown to be due to the *energy* released, proving Einstein's famous equation: $E = mc^2$, in which E is energy, m stands for mass, and c is the speed of light. What this meant in practical terms was that a very small amount of matter could be converted into an enormous amount of energy.

From linear accelerators to cyclotrons, betatrons and synchrotrons, the atom-smashers became ever more powerful. Once it was isolated, by James Chadwick in 1933, the neutron became the favourite 'bullet', because it did not have the disadvantage of an electric charge to repel nuclei, which therefore became easier to penetrate. These machines also produced a bewildering array of *new*, heavy, mainly short-lived sub-atomic particles, one of the first of which was the anti-neutrino. Among others, Murray Gell-Mann in America 'simplified' them into types of *quarks*. Some of them also had a new quality called *strangeness* because they lasted a million times longer than they should – but since their lives should have been only a ten-billionth of a second, they need not worry us here! A final (?) quality of *charm* was also added to explain some 'special' quarks. . . .

In 1939, Otto Hahn and Fritz Strassman tested the action of neutrons on U-235, a then-rare isotope of uranium, and found that instead of disintegrating into unequal pieces, it separated into two almost equal parts – which were found to be barium and the inert gas krypton – emitting some neutrons as it split. More important, it yielded about ten times the energy of previous 'atom-splitting' attempts. It was the first case of *nuclear fission*.

In the US, Italian physicist Enrico Fermi first produced heat energy from the

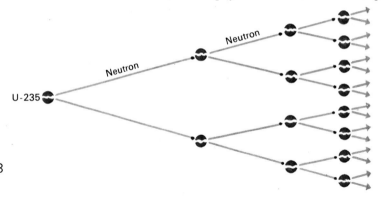

A chain reaction in U-235. The word 'fission' was borrowed from biologists, and indeed we can imagine a uranium nucleus, when hit by a neutron, becoming pear-shaped and finally splitting at the waist, as does a living cell such as an amoeba.

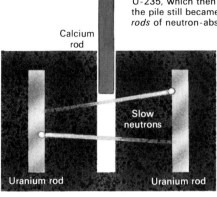
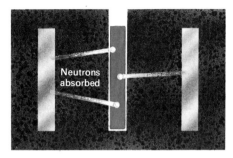

Principle of Fermi's pile. This first nuclear reactor was literally a pile, of 6 tons of graphite blocks set up in a squash court at Chicago University. At intervals were rods of U-238 with a little U-235. The graphite (carbon) acted as a *moderator* to slow down the neutrons emitted by the U-235, which then split U-235 atoms in the other rods. More holes held cooling water pipes, but the pile still became too hot and had to be taken apart, so yet more holes were drilled and *control rods* of neutron-absorbing calcium were pushed in or pulled out as necessary.

uranium reaction on a large scale in 1942. If a piece of U-235 is big enough, each of the two or three neutrons released when one of its atoms is split in half will hit and split *more* nuclei instead of escaping, and so will the ones they hit . . . by the time 50 or so fissions have taken place, billions of atoms will have disintegrated. This is called a *chain reaction*, and the size of a piece of U-235 which will just permit a chain reaction is its *critical mass*. The principle of Fermi's first *pile* is shown here.

Unhappily, the development of the chain reaction occurred during World War II. (Hahn and Strassman were actually working within a few miles of Hitler's headquarters.) In America the vast 'Manhattan Project' was set up, involving 500,000 men and women and two billion dollars, in order to produce the first *atomic bomb*. The previously rare metal uranium had to be mined, refined, and a way found to separate the reactive U-235 from U-238, which was done at Oak Ridge. Meanwhile, at Hanford, Washington, a new, even more fissionable element, plutonium-239, was being produced from U-238 in piles evolved from Fermi's, by way of U-239 and another man-made element, *neptunium*.

On 16th July, 1945, only 31 months after Fermi's first chain reaction, a blinding flash of light, an earth-shaking rumble and a mushroom-shaped cloud over the desert at Alamogordo, New Mexico, signalled to the world the explosion of man's first nuclear weapon. Within a month similar bombs were dropped on the Japanese cities of Hiroshima and Nagasaki, each with the explosive power of over 13,000 tons of the chemical explosive TNT, killing over 100,000 people. On 14th August, 1945, Japan surrendered and World War II ended.

Atoms for war and peace

With the explosion of the first atomic bombs, man had entered the Nuclear Age, for better or worse. Scientists went to work at once on a new type of bomb, so powerful that it needed an atomic bomb just to act as a 'trigger'. At Eniwetok Atoll, in November 1952, the first *hydrogen bomb* was successfully exploded. Whereas the power of an atomic bomb is limited to one *megaton* – i.e. it equals one million tons of TNT – the first hydrogen bomb had a force of 10 megatons. (In 1961 Russia exploded one of 57 megatons.) For while nuclear fission is just that – the dividing of atomic nuclei – the hydrogen reaction *joins* them together. It is nuclear *fusion*.

Fusion requires very high temperatures – millions of degrees Celsius. At these temperatures all the electrons are stripped from the atoms of a light element like hydrogen, forming a *plasma* of free electrons; while their nuclei will move around so fast that they will fuse together should they collide. To break open the nucleus,

Energy and the future

Today's spacecraft use chemical propellants (though nuclear reactors often power their electrical systems). In future, nuclear energy may well be used to reach the outer planets and even the stars. Amongst the many methods proposed are 'ion engines', in which an ionised gas is propelled at immense speeds by an electric field. Here, radiator fins glow red, dissipating excess heat, as a destination is reached.

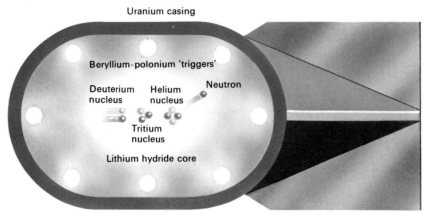

Principle of the hydrogen bomb.

and the strong *binding force* that holds it together, as happens in nuclear fission, takes a lot of energy. But a lot more energy is given out when the nuclei of different atoms are forced together: that's fusion. In a hydrogen bomb, beryllium-polonium cores trigger off plutonium explosions to provide the heat needed to fuse deuterium and tritium (isotopes of hydrogen) to make helium. To increase its power, the bomb's outer casing can be made of natural uranium, which also fissions.

However, one 'advantage' of the hydrogen bomb is that it can be *clean*. A *dirty* fission explosion releases, in addition to deadly radiation, a cloud of radioactive dust known as *fallout*, which causes unpleasant, lingering death from *radiation sickness*. Fusion does not. This would be of little comfort to the few survivors of a nuclear war, perhaps. . . .

But the picture is not all gloom. Today, artificial radioactive isotopes are of great value in medicine and in industry, from the treatment of cancer to the study of wear in machinery (the rate of wear can be found from the amount of *radio-isotope* carried away in lubricating oil). 'Pacemakers' for people with heart trouble can be powered by tiny batteries containing harmless radioisotopes. Vegetables may now be stored for long periods after being *irradiated* with gamma rays from cobalt-60. Nuclear energy powers submarines and will one day propel spacecraft to distant planets.

In nuclear fission, when a proton of hydrogen changes into a neutron in the reverse of beta decay, a new particle – a *positive* electron or *positron* – is given off. During research into *cosmic rays* in 1932, positrons were also discovered coming from outer space.* Where there can be a positive electron (also known as an *anti-electron*), perhaps there could be a whole universe made up completely of *anti-matter*?

We must not forget that all these Laws which have been learned over the centuries are universal. The same atoms, molecules and elements exist in other stars and galaxies and, equally important, our own Sun's immense and constant flow of light, heat and other radiation is produced by the same processes.

The Sun is the key to Earth's energy vaults – which is where this book started. Before we fly further out into space, then, let's come right down to Earth and examine these resources in more detail and see, if there is a crisis, why it came about and what we can do about it. First, we travel back in time a couple of hundred years.

* Amongst other particles detected – adding to the quarks – are the neutrino (which has no mass or charge but has spin and can pass through the Earth!), hyperons, mesons, leptons, muons and pions.

Energy and the future

Another energy crisis?

The year is 1775; the scene, a meeting of industrialists from the rapidly-growing number of factories in Britain. The subject: the 'energy crisis' caused by the fact that suitable sites for waterwheels to power the factories were running out, while at the same time there was a 'timber famine' as wood was used up as fuel, as a

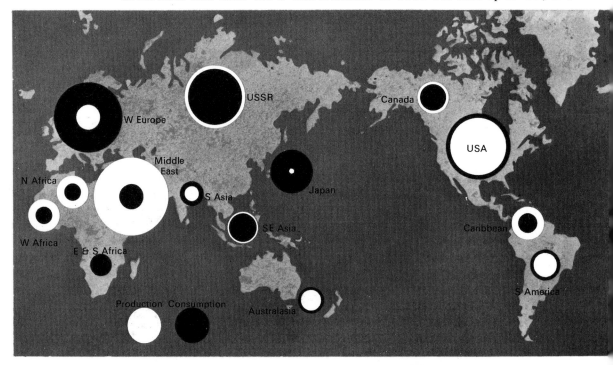

Production and consumption of energy. The amount by which a white disc overlaps black or vice versa shows at a glance by how much supply exceeds demand – or falls short of it.

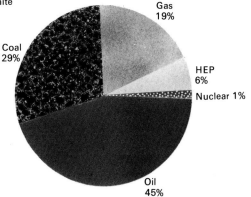

Sources of energy, worldwide (based on 1975 figures). Although oil here accounts for almost half of world consumption, coal has better prospects for the future, while nuclear energy and other sources which do not even appear here will move into prominence.

building material and to make charcoal for iron-smelting. The outlook for 1800, this imaginary conference would conclude, was decidedly gloomy; a limit to growth was being reached. Experts would urge that every effort should be made to conserve energy and to improve the efficiency of waterwheels. (Engineer John Smeaton did in fact double their efficiency.)

The answers to their 'crisis' were in fact already in use: *coal* was used as fuel for *steam engines*, but only close to the coal mines. At first, industry moved away from the chopped-down forests in the south to the coal mines in the north of the country, but the growing canal network later made it economical to transport the 'new' fuel further afield. James Watt's invention could be made mobile, which waterwheels could not, and railways blossomed. A by-product of coal – coke – replaced charcoal in smelting iron ore. For a while all was well.

By 1875 the consumption of coal was increasing so rapidly that to economists another crisis seemed inevitable. Steam engines had improved in efficiency, but a new fuel had to be found for them if growth were to continue. That fuel was *oil*. In turn, oil made it possible to replace the steam engine by new types of motor; between 1867 and 1892 N. A. Otto and Rudolph Diesel introduced the *internal combustion engine* in various forms, but this did not come into its own until the time was right for it. Again, the new engine found uses for which steam could not compete – steam cars have never been a success (although their day may yet come) and there has never been a steam engine-powered aeroplane! As with steam, the full transition took about 60 years, and although of course coal continues to be used, its consumption has dropped while that of oil has risen.

It took first the Yom Kippur War, with the consequent drastic increase in the price of oil from the Middle East to the West in 1973, then the deliberate cut-backs by Iran and Saudi Arabia early in 1979, to awaken most people to the fact that they could not go on running their cars and heating their homes with oil – let alone cheap oil – for ever. Yet another energy crisis had arisen, and although the fortuitous discovery of oil reserves under the North Sea has eased the situation in Britain, it can continue to do so only for a few tens of years. There is a great deal more coal – world deposits of 9 trillion (9×10^{12}) tonnes compared with 220 billion (220×10^9) tonnes of recoverable oil – enough to last for perhaps two centuries. But it is much less convenient than oil in many ways; and if we use more coal to replace dwindling oil, the coal will obviously be used up even faster.

Coal and oil, and the gas associated with either, are known as *primary* or fossil fuels, and it is the doubling in consumption of *these* every 37 years (at a growth rate of 1.9% per year since World War II) that has led to the current 'crisis'. We should take a closer look at these before seeing what alternatives are available.

Energy and the future

How coal was formed. Carboniferous forests were mainly of giant ferns and horsetails. The vertical scale is condensed, as lignite, bituminous coal and anthracite are normally found at depths of 1, 3 and 6 kilometres respectively.
(Above) The process of photosynthesis in a green leaf. In order for the body to use the carbohydrates it must convert them into *glucose* or blood sugar. This is then broken down into carbon dioxide and water with the aid of oxygen from the air – the reverse of the original process by which plants produce sugar. Plants such as sugar beet and cane may in fact be grown intensively in the future to produce alcohol (ethanol) by fermentation as a petrol substitute/additive, and also valuable products such as plastics and chemicals.

The burning rock

Fossil fuels are the Earth's 'energy capital' – like money deposited in the bank years ago (except that we never know *exactly* how much is there). As we saw on page 10, the energy 'income' to build up this capital came mainly from the Sun, whose rays shone down on the primitive swamp plants which covered large parts of the Earth in the Carboniferous period, over 300 million years ago, and on seas which swarmed with tiny organisms. This income *is* still coming in today, but since the primary fuels we 'draw' in one year were deposited over a period of 400,000 years we can ignore any repayment of capital in this form – and there is no chance of an overdraft; once it has gone, it has gone! Fortunately there are other ways of putting this income to work, as we shall see.

Plants containing *chlorophyll* use solar energy to make their 'food', in the form of *carbohydrates* – which consist of carbon, hydrogen and oxygen – from carbon dioxide (in the air) and water, in a process called *photosynthesis*. In doing so they give out oxygen, replenishing that used by all animal life; so although the stored energy of plants *can* be released directly by burning, it would be most unwise to use our forests as fuel. . . . When animals (including humans) eat vegetables, the carbohydrates re-combine with oxygen (oxidise) to give carbon dioxide, water – and energy, which is used by our muscles. The term 'oxidation' applies whether it occurs rapidly when a substance burns in air, when our bodies convert food, or very slowly when vegetable or animal matter *decays* in air, aided by the action of bacteria.

When the Carboniferous forests died and began to decay, their peat-like remains were overlaid with sand, mud and clay. Sometimes new forests grew on these and the process was repeated, but over millions of years the sand and clay, which prevented oxygen from reaching the remains and causing further decay, hardened into rock – sandstone and shale. Thousands of metres of additional rock formed above them, exerting great pressure which squeezed out water and gases and even heated the peat, forming coal. The layers are known as the 'Coal Measures'. Coal, then, may be described as a sedimentary, combustible rock. It is composed mainly of carbon, plus some water, gases and minerals, which are either driven off as smoke and steam or leave a solid residue of ash when coal is burned.

Under the pressure of about one kilometre of rock, the coal formed will be *lignite*, which is often brown, is crumbly and looks woody. There are large deposits in Australia, but little in Britain. Under 3 km, the coal seam will be only half as thick, being compressed into *bituminous* coal – the commonest in Britain. It is black, and splits along definite lines of cleavage, called its 'cleat' by miners. At 6 km, *anthracite* is formed; this is classed as a metamorphic rock, having been changed by heat, and it is hard and shiny. Another type of coal is *cannel* coal, which is hard but dull, and is found in Lancashire and Scotland. Coal seams can vary in thickness

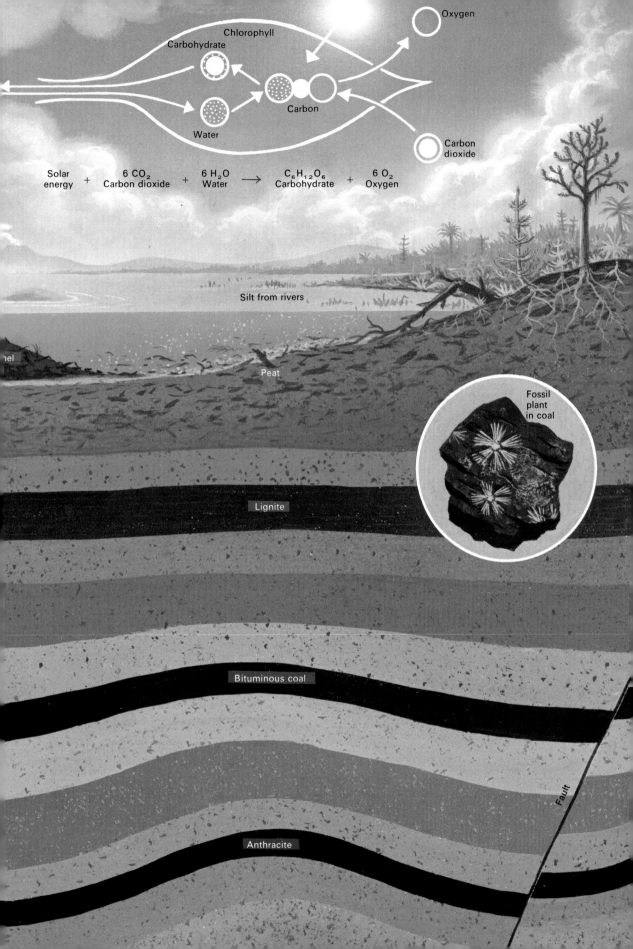

Energy and the future

from a few centimetres to 10 or even 30 metres (though the latter are rare), and are squeezed and tilted by 'earth movements'. Where a seam is badly disturbed it may simply not be worthwhile to try to remove the coal.

Coal for the future

Where the slow but constant shifting of Earth's crust brought coal seams to the surface it is easy to collect it. The first people to do so were probably the ancient Chinese, 3,000 years ago, who raked it from the ground then dug into hillsides to follow the seams. As they burrowed deeper they were liable to be buried as the 'roof' caved in. They, and other early miners, got over this problem by digging vertical well-like holes called 'bell-pits' because of their shape. Hundreds of years later miners learnt how to leave 'pillars' of coal to prop up the roof so that workings could extend several kilometres from the vertical shaft; ventilation then became a problem.

From the beginning of large-scale coal mining in the 16th century, miners worked in terrible conditions, and owners had little regard for their health or safety. This has gradually improved over the years, due largely to the growth of the Trades Unions, until today the mineworkers exert considerable political power. Nowadays machines and 'automation' are widely used, making the miner's work less arduous, and also reducing the number of underground workers. Only half of

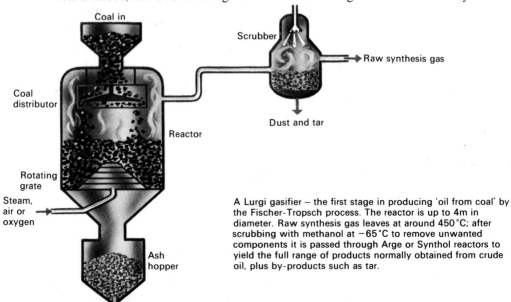

A Lurgi gasifier – the first stage in producing 'oil from coal' by the Fischer-Tropsch process. The reactor is up to 4m in diameter. Raw synthesis gas leaves at around 450°C; after scrubbing with methanol at −65°C to remove unwanted components it is passed through Arge or Synthol reactors to yield the full range of products normally obtained from crude oil, plus by-products such as tar.

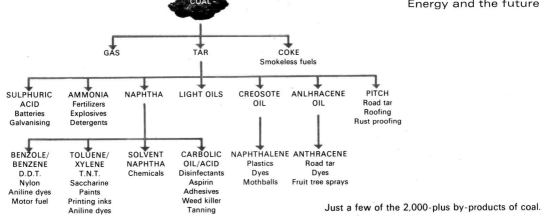

Just a few of the 2,000-plus by-products of coal.

the coal known to be present in any mine can be removed, as the rest is too difficult to reach economically, or has to be left in place to prevent other parts of the mine from collapsing (unless allowed to do so deliberately, as in 'longwall' mining, which permits 60–80% recovery).

Russia has almost 70% of the world's coal reserves, and the USA some 15%, but there are large as yet unexploited deposits in Greenland and Antarctica. Of the 9×10^{12} tonnes of 'in place reserves' (of which only half is recoverable), 76% is of the anthracite and bituminous type and 24% brown coal and lignite. The harder coals produce the most energy, so would appear to be the most economical; being deeper, though, the cost of recovery has to be offset against this. Lignite, on the other hand, can often be obtained by 'strip mining' or 'opencast' methods – but this can scar the landscape badly.

Damage of another kind can be caused by the gases given off by burning coal. For instance, the oxides of sulphur dissolve in rainwater to produce acids, which attack buildings – and lungs; 4,000 people died in the 'smog' of 1952. The Clean Air Act of 1956 declared 'Smokeless Zones' to combat this and the blackening of buildings by soot. Sulphur and sulphuric acid can actually be produced commercially by removing them at source. But one important 'spinoff' from the Act was the development of a variety of 'smokeless fuels'. Anthracite and steam coal are 'clean burners' straight from the pit, but other types have to be made; *coke* is produced by *pyrolysis* – heating coal to around 500°C in the absence of air – yielding many useful by-products.

Although there is about 11 times as much recoverable coal, worldwide, as there is oil – three times as much as all *possible* sources of oil plus natural gas, taken together – simply going back to coal when oil runs out will not solve all our problems. As we saw, coal is of little use for transport; almost the only way it can be used here is via electricity. Over 65% of Britain's electrical energy comes from coal-fired power stations (where most coal goes) at present. In order to feed it to the steam boilers automatically, the coal is pulverised and the powder sprayed in almost like a gas. However, pulverised coal or liquids produced as below could propel road vehicles using spark-assisted diesel engines.

This gives a clue to some other modern (and future) uses of coal: by heating pulverised coal with oxygen and steam to about 1,100°C, *synthesis gas* – a mixture of hydrogen and carbon monoxide – is produced. When passed over a catalyst, a wide variety of oil-type products may be obtained. There are other methods of producing synthetic crude oil or *syncrude*, which will become increasingly important. Some require the addition of hydrogen under pressure and a temperature of 4–500°C to coal 'slurry' (i.e. coal dissolved in a solvent), then filtering, 'cracking' and distillation similar to that used for oil (page 68); this is known as *liquefaction*.

The four most common geological formations in which oil and gas are found (but rarely formed).

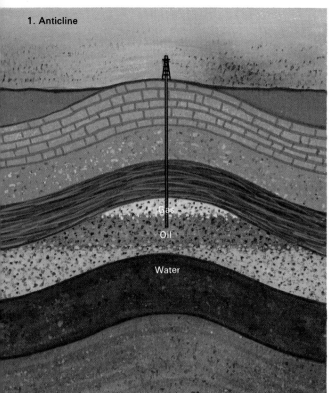

1. Anticline

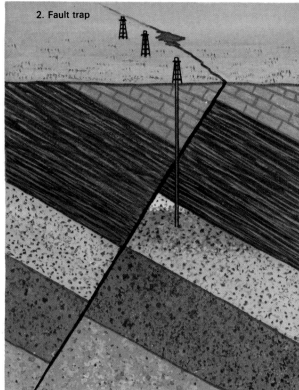

2. Fault trap

From coal to gas

Coke was originally a by-product of Britain's 'town gas' industry, which started in 1810 in London with the Gas Light and Coke Company, but which made way for natural gas after 1967. Being light, coke takes up a lot of storage space, but manufactured smokeless fuels* are usually denser and leave little ash. By a process similar to liquefaction – *gasification* – coal may again yield gas for industrial and domestic use, as *substitute natural gas* (SNG). *Rapid* pyrolysis is more efficient than the old method, but leaves 50–60% of *char* or solid residue (which can, however, be used in other ways). Both coal, which may be of quite poor quality, char, or even domestic refuse can be used in a process called *fluidised bed combustion*, in which small particles of the fuel 'float' on a cushion of air (plus steam if required for gas) blown at exactly the right pressure through the 'bed' on which they burn. If the bed is sufficiently hot almost total combustion is possible; sulphur may be removed from coal by adding crushed limestone or dolomite (carbonates) to the bed.

By immersing steam pipes in a bed, a conventional steam turbine can be operated; or by raising the pressure of the bed the hot gases will drive a gas turbine at the *same time* as the steam is used. About half of the electrical output of the gas turbine is diverted to drive the compressor which supplies the air for the bed. It is hoped that an efficiency of 50% will thus be obtained, against 38% which is the maximum for a conventional steam turbine, and the cost of its energy could be 10% lower. Britain's National Coal Board pioneered this process, but extensive research and development (R & D) is now being done in the US. Such R & D is expensive, but will pay dividends as oil and gas run out.

Meanwhile, what about natural gas? The term can apply to any one, or a mixture

* Eg. Homefire, Sunbrite, Coalite, Rexco and Phurnacite.

The anticline accounts for 80%, the stratigraphic trap for less than 10% of world oil.
Salt domes are common under the North Sea.

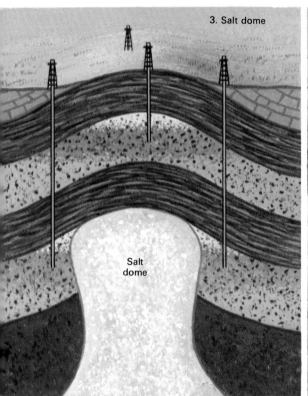

3. Salt dome

4. Stratigraphic trap

of: methane, ethane, propane, butane and pentane. British Gas sends its consumers a constant blend, mainly odourless methane to which a distinctive smell has been added. Methane is often called 'marsh gas'; when found in coal mines, where it can cause an explosion, it is known as 'fire-damp'. The reason natural gas and oil are often mentioned together is that they come from basically the same source.

Gas and oil, like coal, were formed hundreds of millions of years ago – the greatest part in the Jurassic period. Microscopic plants called *phytoplankton*, and tiny animals which ate them, lived in the warm seas. They sank to the bottom as they died, and were covered by layers of silt which later hardened to sandstone and shale. Unlike coal, which stays in the layers in which it was formed (unless shifted by earth movements), the light gas and oil which formed from these organisms under pressure and heat, deep underground, forced their way upwards through porous rocks. Sometimes they reached the surface, forming *seeps*, occasionally still found. However, if they reached impermeable rock a *trap* was formed, especially if the rock had been squeezed into folds. It can be seen that the gas and oil in a field do not necessarily originate from the same place below ground; the gas could have been expressed when coal was formed, for instance, but both arrived eventually at the same place.

Many people still think that oil and gas are found in great underground pools or caverns. Actually they fill the tiny holes or pores in some types of sedimentary rock, under pressure. A crack or fault can easily release them, so that only one in 20 formations which *look* geologically suitable turns out to contain recoverable gas or oil.

Where the gas goes

As we saw earlier, plants use solar energy to produce carbohydrates. These, although composed of only carbon, hydrogen and oxygen (the last two usually in the ratio of 2:1, as in water), form a large group of compounds including sugars, starches, cellulose and gums. The scientific name for natural gas and oil is *paraffinic hydrocarbons*. Hydrocarbons consist of carbon and hydrogen only – which will not surprise you when you remember that the organic matter from which they formed was buried away from oxygen.

'Associated' gas which has formed along with oil is said to be *wet*, and contains fairly large amounts of butane and heavier hydrocarbons in vapour form. *Dry* or 'non-associated' gas consists mainly of methane. In either type, acid gases such as hydrogen sulphide (which smells like bad eggs) and carbon dioxide may be present and have to be removed, usually by dissolving in a solvent liquid which can be re-used after heating. Again, sulphur can be produced, its sale covering the cost of acid gas removal. Water must also be removed before the gas enters the pipeline, to avoid corrosion, while the propane and butane may be extracted from wet gas by cooling and compression. They can then be 'bottled' and sold as *liquefied petroleum gas* (LPG) for cooking and heating (for instance as 'Calorgas'). The remainder is known as *lean gas* or *tail gas*.

There is two-thirds as much natural gas in the world as oil, and future discoveries are expected to equal those of oil. The main gas users are the USA, Europe and Japan, though the latter has to import almost all it uses. About one third of America's total energy needs are met by natural gas; over a million kilometres of pipelines carry this from the wells in the south to the cities of the north east and mid-west, where over 40% of the homes are heated by it (accounting for over half of total US gas consumption). Since 1970, demand has far exceeded new discoveries of gas, leading to urgent examination of sources of supply such as Alaska and the Canadian Arctic islands, or the importing of liquefied natural gas (LNG) or manufacture of SNG. Optimistic estimates suggest that even conventional US sources could last from 35 to 60 years; the lowest estimate is 12.

Until the mid-1960's only some 2% of Europe's total energy was supplied by natural gas, as it was used only fairly close to its source. Then in the late 1950's a huge gas field was discovered at Groningen in Holland, and pipelines from it were added into those already in use by the gas industries of Holland, Belgium, France, Germany and eventually Italy. By 1970 the use of natural gas in Europe had risen seven times, and it now approaches 15% of Europe's total energy consumption. Gas from the North Sea will add to this, but gas piped from Czechoslovakia and Austria, or even from Libya and Algeria by submarine pipeline or as LNG, are possibilities for the future.

We have already seen that natural gas took over from coal gas in Britain from

Energy and the future

1967, when gas cookers and fires in millions of homes began to be 'converted' at great expense. Since then gas has doubled its part in total UK energy usage. The North Sea – or more accurately, the UK Continental Shelf – should continue to meet Britain's needs until well into the 1980's or even 90's, but imports may then become necessary – if they are available.

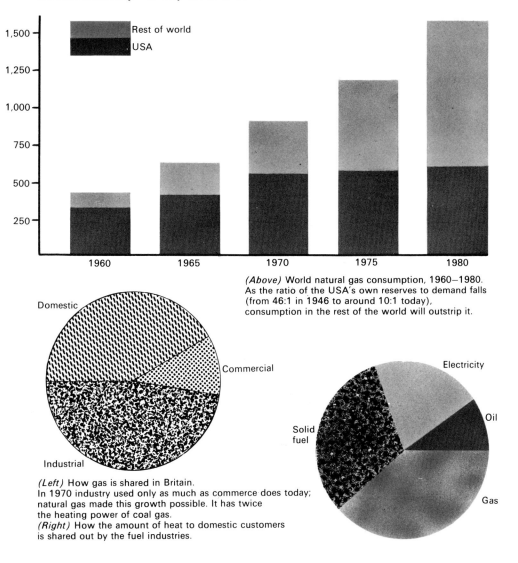

(Above) World natural gas consumption, 1960–1980. As the ratio of the USA's own reserves to demand falls (from 46:1 in 1946 to around 10:1 today), consumption in the rest of the world will outstrip it.

(Left) How gas is shared in Britain. In 1970 industry used only as much as commerce does today; natural gas made this growth possible. It has twice the heating power of coal gas.
(Right) How the amount of heat to domestic customers is shared out by the fuel industries.

Energy and the future

When will the gas go out?

The Chinese were, as usual, first to tap natural gas. They did so as long ago as 1000 BC, using bamboo tubes. Oil seeps were often a nuisance – in Pennsylvania, USA, it ruined good agricultural land. When oil began to be tapped commercially, in the 19th century the associated gas was often burnt off at the well-head. Part of this gas is dissolved in the crude oil and still has to be separated from it to enable the crude to be handled safely. The ratio of gas to oil varies widely, but a typical value would be 200 cubic metres per barrel of crude oil. A 'barrel' is a measure peculiar to the oil industry, and equals 0.136 tonnes (metric) or 159 litres (35 gallons). In the above case the *thermal energy* ('heating power') of the gas would be about one sixth that of the oil.

Since most of the world's reserves of gas and oil are thousands of miles from their consumers, it can be seen that it is much more economical for a pipeline to carry oil than gas. Nonetheless, rising energy prices and advances in technology – such as the development of large diameter, high-pressure pipelines in America – make it increasingly worthwhile to transport gas over long distances. For moving it between continents, though, its *thermal density* needs to be increased. Although expensive, liquefying at source and using special tankers is about the only way to do this.

In associated gas about half of the thermal energy is in the propane and heavier hydrocarbons, from which LPG can be made. These are sometimes also referred to as *natural gas liquids* (NGLs), and may be thought of as intermediates between natural gas and crude oil. To liquefy unassociated gas or the tail gas from NGL plants needs a temperature of $-163°C$, so is very expensive; an LNG plant can cost three times as much as an NGL plant giving the same thermal output. The first sea shipment of LNG to Britain was from Algeria in 1964, when it was used to enrich town gas.

One possibility for reducing these high costs could be to convert natural gas into *methanol* (a 'clean' fuel also known as methyl alcohol or wood spirit) before shipment. One advantage of this is that it is liquid at ordinary temperatures so would

The thermal density of natural gas is only 1/500th that of oil. This means that, for instance, to provide the same quantity of heat a given pipeline would have to carry five times as much gas as oil. As large, high-pressure pipelines are developed, diameters of over 1.5m will not be uncommon, carrying 10^9 cubic metres of gas per day – equal to 25 million tonnes per annum (mta) of crude oil.

Energy and the future

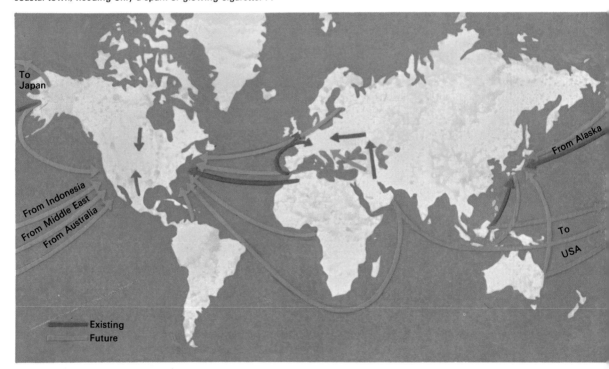

The probable future routes of international gas supply. Improvements in technology of pipeline-laying in deep water could make it possible to transport gas across, say, the Mediterranean. However, for most long-distance movement liquefying and the use of giant tankers will be more usual. Not only is this expensive, but builders will have to bear in mind the implications of a *Torrey Canyon*-type disaster, with heavier-than-air LNG expanding to 1/600th its normal density and flowing into some coastal town, needing only a spark or glowing cigarette...

not need special refrigerated ships; it could be converted back into gas in an SNG plant, though doing so gives a thermal efficiency of only 55% compared with 80% for LNG. However, one other interesting aspect of methanol is that it could be used, either alone or blended with petrol, as a fuel for cars.

The price of gas is governed more by the very large capital investments which have been made in its production and distribution than by the actual cost of getting gas into the system. As cheap 'local' supplies dwindle there will be a strong economic incentive to continue to use the system, even with dearer imported or substitute gas. As we approach the end of the century the search for natural gas will intensify, and as it becomes possible to produce SNG from coal on a large scale gas will doubtless continue in popularity as a clean and efficient prime fuel. In industry, and at power stations in particular, it will gradually be 'phased out' and replaced.

Energy and the future

Black gold

Neither coal, gas nor oil are single substances. We have seen that coal exists in several forms – each suitable for a particular purpose – and that gas too is of several types. 'Oil' is an especially unfortunate term in a way, since to most people it conjures up a picture of a clear, slow-running golden liquid. The more accurate word *petroleum* is not a great deal better, as it is either abbreviated to 'petrol' (called 'gasoline' in the US) to mean petroleum spirit, or brings to mind petroleum jelly ('vaseline'), widely used for medical purposes. Naturally-occurring petroleum or crude oil can range in colour from black – hence the term 'black gold' from the time of the American oil rush – through dark green, browns and reds to pale yellow. In viscosity it varies from very thick – *pitch*, *asphalt* and *bitumen* (said to have been used to waterproof the Ark, and certainly used for thousands of years for surfacing roads, sealing boats etc.), through waxes to very fluid and of course gaseous. Even so, it comes as a surprise to learn that there are literally hundreds of thousands of varieties of these hydrocarbons.

Even where sediments are thick enough we cannot continue to find more oil simply by drilling deeper. Temperature in the crust of the Earth rises with depth, and below about 7,000 m the temperature, at 150°C, is such that the 'cracking' process used in refineries (page 68) operates in nature, but destructively. The hydrocarbon gases, especially methane, are however stable at much higher tem-

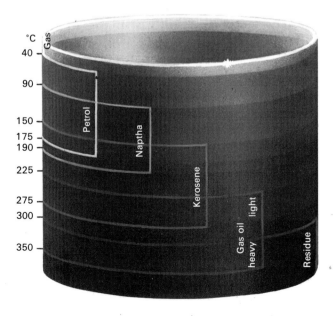

Composition of a typical sample of crude oil, with range of boiling points. The latter increases with the number of carbon atoms in the molecule.

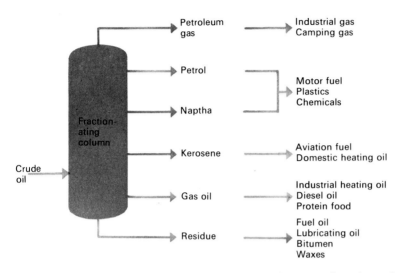

Some by-products of oil. Its importance to our whole way of life is obvious.

peratures, but at great depths and pressures methane occurs as *hydrates*, which are not useable. Another rather startling fact is that in the USA an average of three barrels of brine – water with up to ten times as much dissolved salt as sea water – are produced for every barrel of oil; but the actual proportion of oil to water in a typical reservoir may be only 200 parts in a million, and the rock itself can be ten times the volume of oil and water together!

The area of 800 km by 500 km around the Persian Gulf appears to be unique in the world. It contains half of the world's known reserves of oil. Moreover, the oil in the Arabian fields appears to have been formed actually *within* limestone deposits, which is virtually unknown elsewhere. Yet about half of Earth's land areas lie above sedimentary basins which appear, at first glance, to be suitable for oil production. It now seems certain that there is no 'second Middle East' to come to the rescue, though northern Siberia just might conceal quite large deposits, and quantities perhaps equal to those in the North Sea have recently been discovered in Mexico.

As in the Middle East, oil was found on the surface in parts of America, such as Pennsylvania. At Oil Creek it was scooped up and sold for its healing properties (it still forms a raw material for the pharmaceutical industry). It was in 1855 that a scientist suggested that crude oil could be distilled to make many saleable products, and in August, 1859, Edwin L. Drake *drilled* for what he called 'rock oil' for the first time, at Titusville, Pennsylvania. He struck it at 21 metres on 27th August. Within a day there was a rush akin to the gold rush ten years earlier. At first the main product was *paraffin* (called 'kerosene' in America because paraffin wax was used for candles) for lamps; then came solvents such as naphtha for cleaning and making paints, and lubricants such as cylinder oil and engine oil. Its use as a motor fuel came much later.

Energy and the future

The drilling rig *Orion* flares off gas in the North Sea against a dramatic sunset. Vast quantities of gas were wasted by burning off at oil wells before its potential value was realised, though much is still flared off today.

The costly quest

The first stages in searching for natural gas and oil are very similar. One method is to set off explosive charges, either in the ground or just below the surface of the sea, in places which looked promising when surveyed by aircraft and ship. The reflected waves or 'echoes' from the boundaries of the various rock layers build up a three-dimensional picture, or *seismogram*, on pressure-sensitive instruments. Some £15 million is spent every year on such surveys in UK waters. But only drilling can prove whether oil and gas are actually waiting below!

A drilling *rig* with a tall tower or *derrick* has to be set up. This is quite a hard job on land; imagine the difficulties of doing so in nearly 100 metres of water with waves 20 metres high and winds of over 150 kph. . . . Tackling this has been one of the greatest engineering challenges of the 20th century. The tough and skilled men who work on the rigs, usually 12 hours on, 12 hours off, two weeks at a time then a week off, have names like 'roustabouts' and 'roughnecks'; the latter join together the 'string' of steel pipes lowered from the derrick platform in 10 metre sections. On the lowest end of these a rotary bit chews its way through perhaps three kilometres of rock to the reservoir below. First, though, a steel 'conductor pipe' up to a metre in diameter is positioned in the sea bed with its top above the waves. This prevents water from pouring into the bore hole, and also helps later when special 'mud' is pumped down the inside of the drill pipe to prevent oil or gas from suddenly surging up the pipe. The mud, with added chemicals, also cools and lubricates the bit and carries rock cuttings up the space between drill pipe and bore hole, to the surface.

Even though it may be coated with hard (and costly) industrial diamonds, the bit will wear out and have to be replaced. To change it too soon would be wasteful, but doing so too late would slow down the rate of drilling. All the lengths of pipe have to be pulled up, stacked, then reassembled for drilling to re-start. After all this, only one in 40 such wells succeeds in tapping gas or oil in quantities which are worth exploiting commercially. If it is successful, the next step is to replace the rig by a *production platform* with its legs firmly embedded in the sea floor. From this, wells splay out, not just vertically but reaching to more distant parts of the reservoir. Meanwhile, undersea pipelines have to be laid by a special barge, to be ready to bring the valuable product to its terminal ashore.

Unconventional oil

When gas comes ashore it can usually be piped to the consumer with a fairly small amount of treatment (such as the removal of hydrogen sulphide from 'sour' gas to make 'sweet' gas, which has less than one part in a million of the pollutant) and

Energy and the future

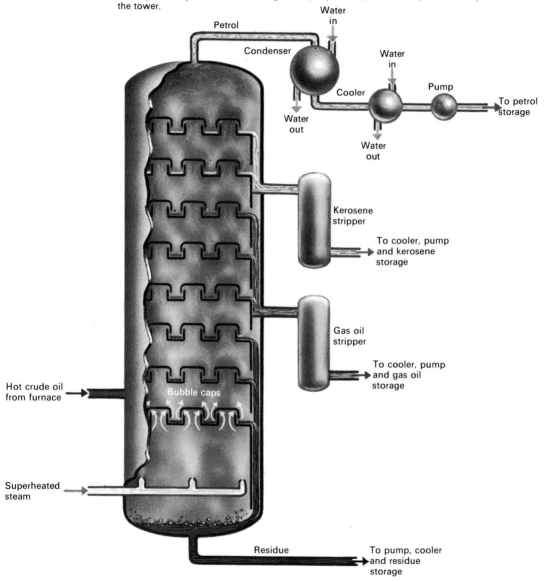

How the fractionating column in an oil refinery works (much simplified). Such a tower may be up to 70m high, and the refinery may have up to 15 of these, connected to 16km of pipes, with 70 storage tanks. Distillation is continuous in trays spaced about 60cm apart all the way up; in each tray are *bubble caps* so that vapour is forced to bubble through the condensed liquid in each tray. Low- and high-boiling materials are thus gradually separated (fractionated) between top and bottom of the tower.

the adjustment of pressure and flow to the demand at any time. The terminal at Bacton in Norfolk, for instance, can deliver up to 1,200 million cubic metres per day. A handful of men are thus in control of the equivalent in energy of twenty 2,000 megawatt (2 GW) power stations – or of burning 1,000 tonnes of coal every ten minutes without stop.

In 'normal' petrol, straight from primary distillation, the molecules are composed of straight chains of atoms as in (1) (n-butane, C_4H_{10}), known as *straight-chain alkanes* or paraffins. High octane motor fuel needs mainly *cyclo-alkanes, branched alkanes* and *aromatics* as in (2) (benzene). These are produced by *catalytic cracking* – splitting the molecular chains of one fraction into a more useful one with the aid of heat and a catalyst (which remains unchanged) to add branches or close loops.

Crude oil, on the other hand, has to be pumped to a *refinery* so that the smelly black liquid can be subjected to *fractional distillation* to break it down into the many products needed in modern life. This is complicated by the fact that every single sample of crude – even one taken only a few hundred metres above, below or alongside its neighbour – is an individual mixture, rather like a fingerprint. The standard unit is the American Petroleum Institute's *degree of specific gravity*; in this, water is given an API rating of 10°. North Sea oils usually fall into the range 35–45°, which are classed as 'light',* but worldwide, crudes as heavy as 5° and as light as 60° have been found. North Sea oil is also 'sweet' in terms of sulphur content (0.2–0.4% by weight), while Middle East and Mexican crudes are 'sour' – enabling over 60% of world sulphur needs to be recovered from oil and natural gas.

The diagram makes the processes of a refinery clearer than words. Boiling points increase with increasing molecular weights, from $-10°C$ for the gases to $350°C$ for diesel and fuel oils.

When all the continents and their undersea shelves, including those in Arctic regions, have been fully explored and exploited, and the oil in well after well dries up – what then? Oil will not be finished; although only some 30% of the crude in any reservoir can be extracted by conventional means, much research is being done in an attempt to increase this to 70%. Some 'secondary recovery procedures' are to pump gas or (possibly hot) water, steam or chemicals back into the reservoir rock under pressure, with the aim of lowering the oil's viscosity; or explosives may even be used.

Meanwhile, sources of oil which have previously been ignored as being 'uneconomical' begin to look attractive. *Tar-sands* or *oil-sands*, such as found in Athabasca (Alberta) in Canada and in Venezuela, are estimated to contain $1,500 \times 10^9$ barrels of oil, removable by the 'secondary' methods above or even by opencast mining. Ten syncrude plants, costing over £500 million each, could produce $1\frac{1}{4}$ million barrels of oil a day from the Athabascan oil-sands by 1985, but its low gravity, high viscosity and sulphur content mean that it needs upgrading to a lighter oil.

Oil-shale, in which oil exists as a solid called *kerogen*, which decomposes at about 370°C to give a light 'shale-oil', is found in massive amounts in Colorado – $6,850 \times 10^9$ barrels, more than all Middle East reserves together! An unusual method of using oil-shale could be 'retorting' the hot vapours on the spot, in natural fissures or in caverns blasted out of the shale, but opencast or old-fashioned 'room-and-pillar' deep mining seems more likely. Before oil-shale becomes profitable we have to be sure that the energy consumed by mining, transport and heat to extract and refine the oil is well below that produced by the final product!

* Making it suitable for petrol, but meaning that the UK must exchange some for heavier oil or sell it in order to import the latter, even after 1980 when the North Sea could otherwise supply all UK needs.

Energy and the future

Fossil fuel reserves. If the annual increase in energy demand continued at its present rate, a total of almost 20 Q would be required by the end of the century, mainly as oil or natural gas.

COAL	200 Q
OIL	12 Q
NATURAL GAS	12 Q
OIL SHALES	25 Q
TAR SANDS	7 Q
TOTAL	**256 Q**

The nuclear alternative

My dictionary defines 'crisis' not as a disaster but as: 'the point when an affair must soon undergo a change for better or worse; a period when momentous changes are effected'. That certainly seems a fair description of the situation where energy is concerned. Let's pause to look at how we use energy at present.

Because the quantities involved are so colossal, it is usual to describe energy demand in terms of the unit 'Q'. One Q is 10^{18} British thermal units (the Btu as a unit of heat has, as we saw, been largely replaced by the joule in the modern SI system, but equals 252 calories – a calorie being equivalent to 4.2 joules). More to the point, 1 Q is equal to about 200 billion barrels (25×10^9 tonnes) of oil. This makes the world's annual consumption of energy in all forms $\frac{1}{4}$ Q, increasing by 7% each year. The reserves (which means *proven* recoverable amounts, as opposed to 'resources', which *may* be exploitable in the future) of fossil fuels are shown above. Remember that this is all 'capital'. The *annual* 'income' of solar energy used in photosynthesis by plants is about 1.2 Q, of which we use about 7% as food and by burning.

It is clear that, although coal at least *could* last until AD 2200 with care, there has to be a 'change for better or worse' before the end of this century. Gasification and liquefaction of coal will aid the changeover; but what are we changing *to*? Is there any other form of energy capital waiting to be used, or do we need to turn to income? As with those earlier 'crises' on page 53, the answers are already with us – but never in history has there been such controversy over the alternatives. The obvious winner – but in some quarters the least popular – is *nuclear energy*.

Every energy source has its attendant problems. Coal miners are killed, coal itself pollutes the atmosphere and soil when burnt; oil can escape into our oceans, causing huge slicks, and refineries and natural gas tanks can explode. Perhaps because nuclear bombs wreaked such havoc during World War II, and the after-effects on their human victims were so horrific, the public feels an understandable suspicion about nuclear power. Is it justified?

There is no mystery about a nuclear power station. The heat to produce steam for the turbo-generators comes from a nuclear reactor instead of from burning coal or oil in a furnace; the electricity which leaves the station is exactly the same. Present nuclear reactors use fission, which means that the heavy atoms in the fuel – usually uranium – are 'split'. The fuel is surrounded by a *moderator*, which may be graphite, water, or 'heavy water', to slow down the high speeds of neutrons which fly out from the nucleus of the atom which gives nuclear energy its name. Following prototypes built at Windscale in 1950, Britain commissioned the world's first commercial nuclear power stations in 1962, and there are now nine *Magnox* 'advanced gas-cooled reactors' (AGR) stations with graphite moderators, using a magnesium alloy for 'canning' uranium fuel elements, in service. Britain considers the AGR safer than the cheaper American LWR (light water reactor),* though this has sold well.

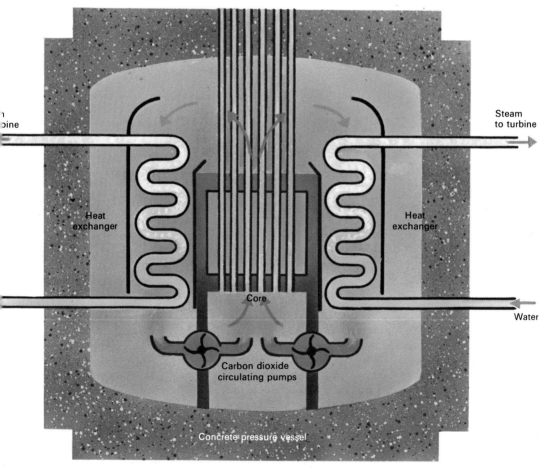

Schematic view of a Magnox-type advanced gas-cooled reactor.

---

* An attitude which many feel was justified by the narrowly-averted disaster in the pressurised water reactor (PWR – a form of LWR) at Harrisburg, Pennsylvania, in April 1979.

Energy and the future

Problems, problems . . .

The record of nuclear power in Britain, so far at least, is a good one. The stations produce 12% of Britain's electricity, more cheaply than coal- or oil-fired stations, and are considered safe and reliable. But are they really efficient? Current reactors are wasteful of uranium, consuming only about 1% of that mined. Reserves of uranium amount to 1 Q as used at present, but a new type of reactor, the *fast breeder*, could extend this to 150 Q. This is because present reactors 'burn' the rare U-235 isotype of uranium, of which only 0.7% exists in natural uranium or U-238. A fast reactor is able to convert abundant and cheap U-238 into plutonium, then burn this as a fuel. All reactors do this to some extent, but the fast reactor has a

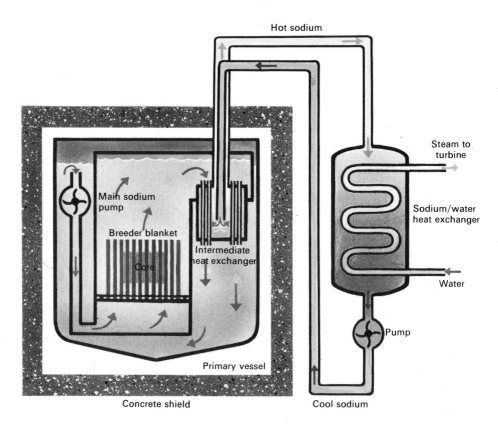

Principle of the fast breeder reactor ('fast' because the neutrons are not slowed down). Its proponents point out that while 1 tonne of uranium in a Magnox reactor will produce 3×10^7 units of electricity (and 1 tonne of coal in a conventional power station 2,400 units), 1 tonne of uranium converted to plutonium in a fast breeder will deliver 48×10^8 units.

better conversion ratio which makes it not only self-sufficient, but it can actually create *more* fuel from U-238 than it consumes – hence the term 'breeder'. It can also convert 'depleted' uranium from conventional reactors into more plutonium. Britain (which has to import all its uranium) now has some 20,000 tonnes of depleted uranium in stock, which in fast reactors could produce the energy equivalent of 40×10^9 tonnes of oil – over five times as much as the North Sea is expected to yield. Instead of requiring half a million tonnes of uranium oxide ore every 15 or 20 years, this amount could suffice for our electrical energy needs for several hundred years, in breeders.

This sounds very encouraging; so what are the problems? We should establish right away that no nuclear reactor can explode like an atomic bomb; but there is no doubt that the fissile materials used in a power station *could* be turned into weapons, if they got into the wrong hands. Stringent precautions have to be taken to ensure that terrorists and other hostile parties cannot divert fuel being transported from place to place. Plutonium is a very dangerous chemical – not only is it extremely poisonous, but being radioactive the tiniest particle can produce cancers, especially if inhaled. The main problem, though, is waste. Radioactive waste materials are produced at all stages of obtaining and using nuclear fuels, but the most difficult are the used fuel elements from a reactor. When these arrive at the reprocessing sites, long-lived radioactive wastes have to be separated from re-useable fuel by being dissolved in nitric acid; uranium (97%) and plutonium (1%) are removed from this. The remaining acid has to be disposed of, but the few percent of wastes it contains remain active for *thousands* of years. This means that we could be handing the problem down to our children's children's children. . . .

At present the wastes are stored in double-walled stainless steel tanks inside concrete cells, water-cooled because radioactivity produces heat, and British Nuclear Fuels Ltd. even stores the waste liquids left after it has reprocessed irradiated fuel from reactors in other countries such as Japan, earning money by doing so. In 1977 a public enquiry took place into the proposed expansion of BNFL's reprocessing plant at Windscale, with bitter arguments for and against. BNFL pointed out that if spent fuel is not reprocessed all the energy in the plutonium and unused uranium is lost. The objectors claimed that toxic releases into the air, sea and onto agricultural land are inevitable, and that only a 'police state' could cope with the threat of diversion, sabotage or terrorism.

The answer most likely to be accepted internationally is one proposed by a committee of the National Academy of Sciences in the US in 1957. This is to bury such wastes in deep, geologically stable rock formations – perhaps in salt deposits. Additionally, the wastes could be turned into glass-like solids by heating with silica and borax; BNFL plan to bring a test plant into operation in the 1980's.

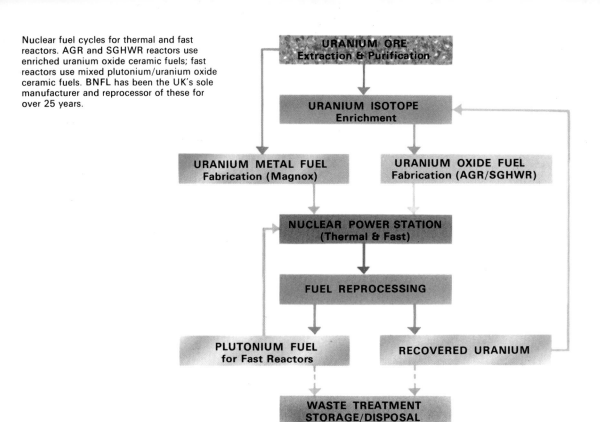

Nuclear fuel cycles for thermal and fast reactors. AGR and SGHWR reactors use enriched uranium oxide ceramic fuels; fast reactors use mixed plutonium/uranium oxide ceramic fuels. BNFL has been the UK's sole manufacturer and reprocessor of these for over 25 years.

Fishing for fusion

I have been referring to the problems of the United Kingdom, but the situation is much the same in other countries. During 1977 protesters died in demonstrations against fast breeder reactors in France, and many American states have strong opposing factions. Even the 'underground burial' solution raises the question of what would happen if the glassy solids began to break up (the maximum temperature, in rock, would be reached 40 years after burial); it would be difficult to retrieve them. Against this, we should not forget that while radioactive waste remains hazardous for a very long time, many 'everyday' chemicals – such as arsenic and barium – are poisonous forever.

The UK's first experimental fast reactor was commissioned in 1959 at Dounreay in Scotland, came to full power in 1963 and was closed down in March 1977, its purpose fulfilled. Work went ahead on the 250 megawatt Prototype Fast Reactor (PFR) at Dounreay, which was commissioned in 1974. Using the experience gained with this, it would take 2–3 years to commence work on a full-scale fast reactor, plus perhaps eight years to build and two to commission it, at a total cost which could be £1,000 million. It could thus be at least 1990 before the first comes into operation, given Governmental go-ahead; another public enquiry has to be held first.

There is not room here to describe all the alternative types of reactor being developed, such as SGHWR (steam-generating heavy water reactor: the hydrogen – *deuterium* – in 'heavy water' contains an extra neutron); or CANDU (a Canadian-designed slow reactor which uses heavy water as moderator *and* coolant), though, using ordinary 'cheap' U-238, this has several advantages. One of these is that it can be adapted to use thorium-232 with the U-238, producing the isotope U-233 instead of U-235, and thus foiling potential bomb-makers; another is that it offers an acceptable substitute for the fast breeder. Clearly it will always be necessary to site nuclear power stations away from the centres of population which use their electricity; Professor Peter Kapitza of Moscow has even suggested building them on small, uninhabited islands. All nuclear stations need cold water from rivers or the sea for cooling exhaust steam back to water.

Indeed, oceans feature more and more in energy 'scenarios' for the future. The greatest hope according to many scientists, if it can be realised, is fusion power which, as we saw on page 50, was used in the hydrogen bomb. It is one thing to make a bomb, but quite another to *control* the forces – and temperatures of 100 million °C – involved. The idea seems simple: make two deuterium atoms combine and they either make a helium atom and release energy, or they make one atom of *tritium* (hydrogen with two extra neutrons in its nucleus) and one of normal hydrogen, also releasing energy. One molecule of water in 5,000 is heavy water, so there is no shortage of 'fuel' in our oceans; if all the deuterium in 1.5 cubic kilometres of sea-water could be used it would release energy equal to the world's crude oil reserves. And there are about 1.5×10^9 cubic kilometres of sea-water available... ($= 10^{10}$ Q). There are about 30 other possible fusion reactions, some using the metal *lithium*. None give long-lived radioactive wastes, and fusion should be 100 times safer than fission.

Experiments with 'Tokamak', a test reactor sponsored by America's Energy Research & Development Administration (ERDA), which uses a 'magnetic bottle' invented in Russia to confine the immensely hot plasma of ionised deuterium gas, and earlier British ZETA, have solved some problems but raised others. Another suggested method is to use intense laser beams to force the two light elements together. But despite intense R & D in the UK, USA and USSR, fusion power is not likely to affect world energy supplies for 50 years.

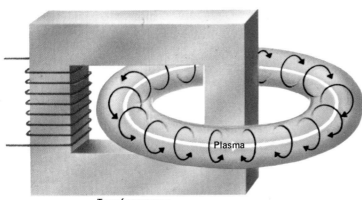

Transformer core

To contain the hot plasma in a fusion reactor such as a tokamak, a hollow doughnut-shaped 'pinch tube' is used. Current through the winding on the transformer core creates a strong induced current in the plasma, whose own magnetic field then pulls it in, away from the tube walls. 350 million degrees C has to be achieved to make the thermonuclear reaction self-sustaining; but as the density of the plasma may be only 1/10,000th of an atmosphere the amount of 'heat' is very small. Far from vaporizing the walls, as is often stated, the kinetic energy of the gas particles would be instantly damped by contact with them! To produce power, fast neutrons emerging from the fusion reaction would be trapped and the resultant heat fed to a steam turbine/generator in the normal way. The Princeton Large Torus has so far achieved 60 million degrees.

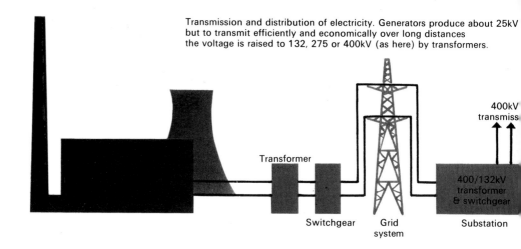

Transmission and distribution of electricity. Generators produce about 25kV but to transmit efficiently and economically over long distances the voltage is raised to 132, 275 or 400kV (as here) by transformers.

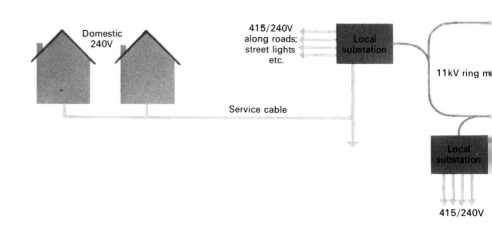

Electricity for all

Oil took over from coal, coal may make a comeback but nuclear fission is making a takeover bid while fusion glows on the horizon, with only 'instant gas' a serious contender for the domestic market. Most of these fuels are converted into electricity of course; would it not make sense to go 'all-electric' *now*, since we may have to eventually, using whatever fuel is available for our power stations at any one time?

The idea does have attractions, but there are dangers in putting all one's eggs in one basket (suppose all power station engineers went on strike). Electricity is actually wasteful as an energy form. Producing it consumes three or four times as much primary energy as it delivers as electrical energy – but would you be without the convenience of light at the click of a switch? What other form of energy can operate your radio, TV, stereo, washing machine . . .? Even so, the required growth rate for electricity production in Britain could be *halved* if electricity were used only where electricity *must* be used, and if energy were conserved wherever possible. Warm waste water and air from power stations could be put to use. The same must apply to a greater or lesser extent in other countries.

This seems a good moment to pause to look at electrical production and distribution. Electricity has come a long way since Joseph Henry and Michael Faraday independently discovered electromagnetic induction around 1830 (page 37), leading to today's dynamo or generator and the electric motor. The first generator to be used commercially was made by a Belgian, Z. T. Gramme, in 1871, and in 1882

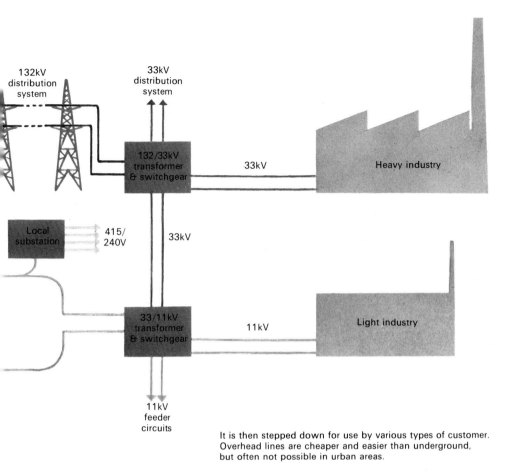

It is then stepped down for use by various types of customer. Overhead lines are cheaper and easier than underground, but often not possible in urban areas.

the first steam power stations in Britain supplied public lighting as well as the private consumer from Holborn Viaduct (at 110 volts DC). Between 1904 and 1914 advanced power stations were built, especially at Carville, Newcastle upon Tyne, and in 1926 the Electricity (Supply) Act provided for the creation of a public corporation – the Central Electricity Board – to interconnect a number of regional stations, forming a national *Grid*.

Voltage in Britain was standardised at 240 V AC in 1947. Also in 1947 the production of turbo-alternators was standardised at two sizes: 30 MW and 60 MW (page 19). A power station output of 1 GWh is equivalent to about 135 tonnes of coal, or 80 tonnes of oil, or nearly a million cubic metres of natural gas (but these are not the amounts needed to *generate* 1 GWh of electricity – remember that the process wastes a lot more of these than it uses).

Today the CEGB has 168 power stations with 785 generators and a combined output capacity of over 58 GW. England and Wales are divided into seven Grid Control Areas, each with its control centre but supervised and co-ordinated by the National Control Centre in London, which ensures that electricity is always available where and when it is needed. National Control also provides for transfers from the South of Scotland Electricity Board and Electricité de France, and so the link-up goes. The Boards also ensure that the siting of new stations and transmission lines cause a minimum of damage to the environment. Similar organisations exist all over the world.

Annual energy received, transformed and re-radiated into space by the Earth. Left to itself, an almost exact balance is maintained; even the 1.2 Q per year stored as chemical energy by vegetation is eventually returned to the atmosphere as heat when plants decay. Only man upsets this equilibrium, by accelerating the release of heat from fossil (and nuclear) fuels and releasing carbon dioxide from the former, trapping heat from the surface of the Earth....

Are we cool?

Now I used to think that I was cool,
running around on fossil fuel,
Until I saw what I was doin'
was driving down the road to ruin.

These words from a James Taylor album have a lot of truth in them. Is it really sensible to use up all our valuable reserves of primary fuels, particularly oil, from which so many products can be synthesised – from 'feedstock' for the chemical industry (resulting in plastics, resins, fibres, solvents, detergents and a host of other items, many of which will be very difficult to produce in any other way) to protein for animal or even human consumption – simply to heat our homes or allow us to drive around in cars? Cars, incidentally, are great energy wasters; usually underoccupied and over-powered, their efficiency is often as low as 10%.

Yet nuclear power is obviously not a perfect answer, no matter what safeguards are taken. Is there any safe alternative energy source which does not pollute, has no hazardous waste products to be disposed of, and is of no use to terrorists etc.?

The answer is 'yes' – and furthermore, since the 'oil crisis' of 1973 an almost bewildering array of solutions has come to light. This in itself almost constitutes a problem, for governments have to decide where to allocate R & D funds, and if in doubt are liable to turn to the power source pushed most often in front of them – such as the fast reactor. Britain's budget for R & D into what are variously described as 'alternative', 'renewable', 'unconventional', 'non-depleting' or 'ambient' energy sources, allied with 'appropriate technology', was recently increased to £10 million, which is a drop in the ocean compared with nuclear research.

The basic source of most of this alternative energy has been around for longer than any other. It is in fact a fusion reactor – 1,400,000 km in diameter, 150 million km away, in which millions of tonnes of hydrogen are continuously converted into helium, the balance leaving as energy. The *Sun* has already featured in our energy story as providing the capital stored in our fossil fuel; now it seems to many people that the time has come to use it as income.

The Sun's rays reaching Earth each *year* amount to 4,000Q – 16,000 times as much energy as we use in a year. This means that, in theory, an area one tenth that of Arizona could supply the electrical needs of the USA. Even in Britain (according to the International Solar Energy Society), if 1% of our land area could convert solar energy to electricity at 10% efficiency, the whole of our electricity needs could be met; and from May to September the average solar radiation received by the UK is as good as anywhere else. One disadvantage of solar power is that most sunlight falls when (summer) and where it is least needed! But the solar rays falling on a square metre of Earth's surface can equal the energy of a one-bar electric fire. So how can we harness all this free energy?

Two 8 m² panels on the author's roof provide all the domestic hot water needed on a reasonable summer's day. Even in winter the temperature of cold water entering the tank is raised sufficiently to save a considerable amount of conventional energy otherwise needed. The two basic systems are *direct* and *indirect*; in the former the water flowing through the panels will eventually come out of the taps, via a 'solar tank'. In an indirect system the sun-heated water re-circulates, becoming hotter, and feeds a heat exchanger connected to the hot water tank.

Powerhouse in the sky

Archimedes is said to have focussed the Sun's rays on the Roman fleet at Syracuse with metal mirrors in 212 BC, setting it on fire. Many uses for the Sun have been found since then; the same principle is used in the huge solar furnace at Mont-Louis in the French Pyrenees, in which 3,500 small mirrors can produce a temperature of 3,000°C – enough to melt many metals. Such an arrangement can be focussed onto a boiler to produce steam for heating or to drive a turbine, though it needs to be steered to follow the Sun's passage through the sky. A flat 'tracking mirror', driven by an electric motor and called a *heliostat* can ensure this. Taking this a stage further, the Sandia Laboratories in America have built a 'test facility' consisting of a boiler mounted on top of a 60-metre tower, surrounded by over 300 large mirror modules turned to the Sun by 78 heliostats. The ultimate aim is to produce a 10 MW solar power station – enough for a town of 10,000 people – by 1980. More advanced concepts involve 20,000–30,000 mirrors.

The only solar collector so far at all widely used domestically provides not electricity but hot water. You have only to place your hand on the roof of a car on a fairly sunny day to realise the heating power of *infra-red* or long wave radiation*; a well-designed 'flat-plate collector', consisting basically of glass-covered black-painted copper panels through which water is pumped, as shown here, can halve domestic hot water costs even in Britain (where an average of 1,500 hours of sunshine is received each year, and other heating costs are rising sharply).

The idea was used 25 years ago in Israel, where ordinary radiator panels were painted black and mounted on roofs, and even earlier in America. It is also especially suitable for heating swimming pools; domestic space heating (i.e. 'central heating') is more difficult at present, but the Solar Energy Consortium in particular have designed a complete domestic installation, now in use in bungalows at Newark, Notts. These include a *heat pump* system which extracts heat from the atmosphere, like a refrigerator, by means of a liquid which is compressed to produce a gas at a much higher temperature (42°C) and then transferred by another 'heat exchanger' inside the building to heat its interior. This has the advantage of providing a cold store in an outhouse at the same time. It is however quite possible to take heat from a swimming pool, stored in the water heated by the Sun, and use it to heat a building – even an office block – via a heat pump; or even to extract heat from the ground.

* See *Light and Sight* by David A. Hardy (World's Work, 1977).

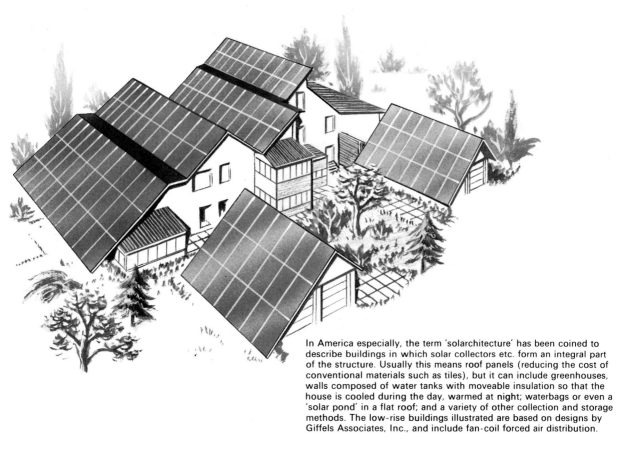

In America especially, the term 'solarchitecture' has been coined to describe buildings in which solar collectors etc. form an integral part of the structure. Usually this means roof panels (reducing the cost of conventional materials such as tiles), but it can include greenhouses, walls composed of water tanks with moveable insulation so that the house is cooled during the day, warmed at night; waterbags or even a 'solar pond' in a flat roof; and a variety of other collection and storage methods. The low-rise buildings illustrated are based on designs by Giffels Associates, Inc., and include fan-coil forced air distribution.

Domestic space heating consumes 17% of the UK's primary energy needs at present. Solar water heaters alone could save up to 2.5% of the total, according to the Department of Energy's monitoring Energy Technology Support Unit (ETSU) at Harwell.

Converting solar energy directly into electricity is very expensive today, but available methods do have the advantage of long life, a minimum of maintenance and no moving parts. (In other words, almost all the cost is in materials and installation – thereafter, as with heating panels, they are 'free'.) The solar panels on spacecraft use silicon cells refined from sand, which convert only some 10% of solar to electrical energy and would have to be about 300 times cheaper to be economical for general use; but experiments in progress with gallium arsenide, metal-and-glass panels dipped in copper solution, silicon crystals pulled from a molten ribbon, and new 'semiconducting glasses' are all aimed at lowering costs greatly.

Energy on the beam

The Sun, as we all know, does not shine on us all the time. And it is just when it is not shining – on dull days, in winter, at night – that heat, or electricity to produce heat or light, is likely to be needed. A solar water-heating system can easily be 'backed up' by conventional fuels, and if solar/electrical conversion ever takes place on a large scale the same will apply, at least for some years. One problem with electricity is that it is difficult to store it – one cannot pour it into a tank like petrol, to be there until needed. Batteries have of course existed for many years, ever since Volta's first 'Voltaic pile', in fact. The ordinary dry cell' or Leclanche zinc-carbon battery is of limited power and life, and has been replaced by alkaline manganese for some applications; while the traditional lead–acid rechargeable

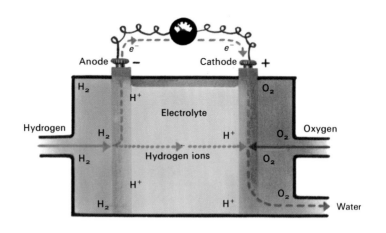

A fuel cell. Each gas penetrates only part-way into its porous catalytic electrode plate, but just sufficient diffuses through to the side wetted with electrolyte to form a thin layer – not gaseous, but *adsorbed* on to the surface. Hydrogen ions (H^+) formed at the anode give up electrons (e^-), which travel along the outside circuit via, here, a voltmeter, to the cathode (NB. anode is negative, cathode positive in a fuel cell). Here, oxygen combines with electrons and hydrogen ions, forming H_2O. A single 'ideal' fuel cell would operate at 1.23V, whatever amount of current was drawn – a more efficient use of hydrogen than in a heat engine. In practice, 'blotter' containing electrolyte is sandwiched between plates in a 'stack' of perhaps 500 cells, each giving 500W at .65V.

'secondary' battery as used in motor vehicles and the newer nickel–cadmium version (both give out 75–80% of the energy put in) will also give way to more exotic types. The slow-moving milk van illustrates the limitations in applying electric propulsion to vehicles; one solution under development may be the sodium–sulphur battery, which has to operate at 300–350°C, but much work needs to be done.

When America's space programme got under way it was widely suggested that the *fuel cells* used to supply both electrical power and drinking water for the astronauts could provide the answer to electric propulsion. In these, hydrogen and oxygen are combined, through a 'diffusion membrane', into electricity, water and heat – in effect, the reverse of electrolysis (pages 31–33). A novel development of this by Battelle Laboratories in America electrolyses water during 'charging' and reconverts the gases into water on discharging. It is claimed to be 50% efficient, and its inventors hope for a 20-year cell life.

It is in space that the most exciting possibilities for solar energy generation lie. A growing number of people believe that one of the future functions of America's new Space Shuttle should be to ferry into low orbit the raw materials for a satellite solar power station (SSPS), to be boosted further out by an ion rocket stage. In a 'synchronous' orbit (turning with the Earth) a satellite receives almost continuous sunlight, and six to ten times as much energy as could be collected by a system of the same size on the ground. 35,000 km out in space, the solar energy would be converted into a *microwave beam* to be received by a ground station on Earth, unaffected by day/night or cloud – and no heliostats needed. The transmitting antenna could be as much as a kilometre in diameter – but the solar cell arrays might be 20 by 5 km. This poses no great problem in airless, weightless space.

Both Boeing and Grumman have detailed designs for such stations, in the 4–15 GW range. The famous science fiction and fact writer Isaac Asimov has written an open letter to President Carter stating that he believes solar power stations, leading to larger inhabited space stations (perhaps 'colonies', as envisaged by the L-5 Society) may be 'the only road that will lead to the salvation of civilisation'.

A full-sized SSPS nearing completion. Behind it is a 1.7km diameter 'Stanford torus' type space colony, many of whose inhabitants built the power station, using raw materials ferried from the Moon or asteroids. At each end of the 100km² power satellite is an antenna which beams microwaves, continuously facing Earth although the solar panels always face the Sun (diagram *left*). On reaching Earth's surface the microwaves are converted into electricity at 90% efficiency by a *rectenna* (rectifying antenna) covering several square kilometres but of 'chickenwire' construction, shielding the ground beneath and allowing crop growth or even grazing. (Nuclear plants would require a similar area for cooling ponds.) Possible damage to living tissue and heating effects on the atmosphere by microwaves are being investigated. An alternative transmission method is the use of infra-red laser beams.

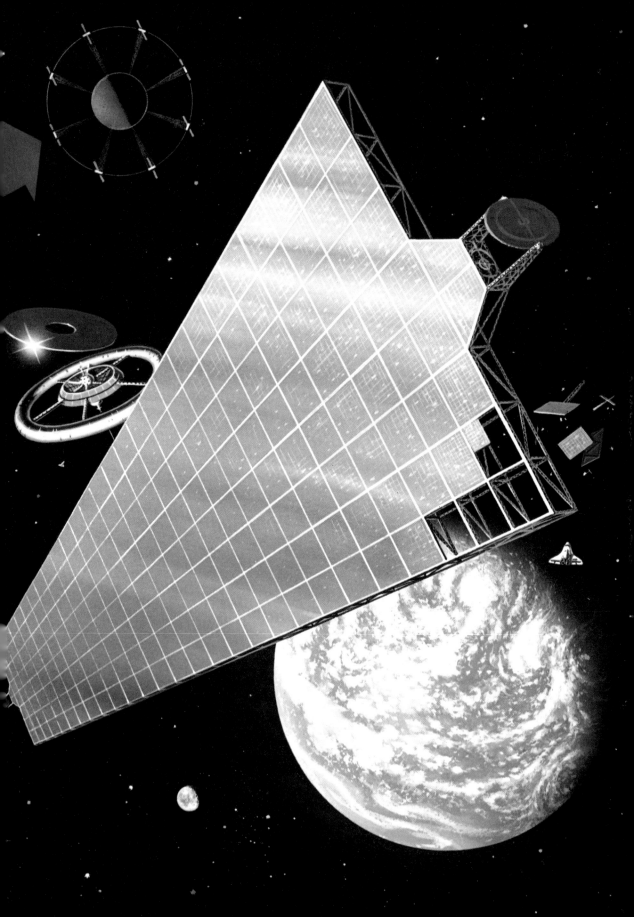

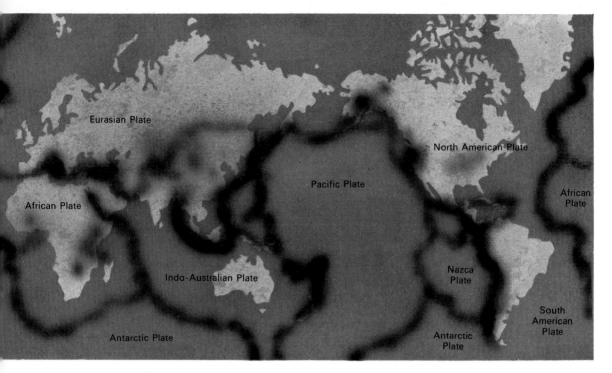

Earth power

Solar energy can be harnessed in a number of very different ways. Before we look into these, though, there is one other form of available, almost non-polluting energy (which may be seen as income *or* capital), which is not dependent upon the Sun, for it comes from inside our own planet.

Geothermal energy has been known to Man for a very long time. In countries such as Iceland and New Zealand hot water bubbling up from the rocks has been used for cooking and other domestic purposes from the earliest days. Volcanoes are evidence enough that great amounts of energy are stored inside the Earth; rather than being due to heat left over from the time when the planet was molten, geothermal activity may be allied to nuclear energy, since it is probably produced by slow 'decay' of radioactive elements deep inside the Earth, but most is due to the meeting of continental 'plates' as in the diagram.

The earliest use of natural steam to produce electricity was in Lardarello in Tuscany, Italy in 1904, where generators driven by steam from wells drilled for the purpose now produce almost 400 MW. There are also geothermal plants in Japan, New Zealand, Iceland (where the capital city, Reykjavik, is almost completely heated by geothermal wells), Russia and America. In a Californian valley called the Geysers – thought by explorer William Bell Elliott to be the gates of Hell when he first saw its erupting jets of steam in 1847! – an attempt to tap the steam source failed in 1922 due to the corrosive effects of impurities (which include hydrogen sulphide) on turbines and pipes. Another attempt in 1950 blew out the top of a hill. By 1960, new stainless steel alloys allowed 11 MW to be produced, with a target of 900 MW before 1980.

Worldwide, over 1 GW of electricity is now produced from geothermal sources. There are two main classes: *dry fields*, in which steam emerges naturally at low pressure and may also be used to produce fresh water by condensation; and *wet fields* (which may be 20 times more abundant), which contain hot water above its boiling point at atmospheric pressure. This does not become steam until the pressure is released by drilling, when hot water plus 10–20% steam spurts forth. A third type, the *low-temperature field*, is now receiving attention. This consists of a large

Geothermal activity — especially volcanoes — occurs mainly along definite lines of weakness in Earth's crust. The principal zone is the 'ring of fire' around the Pacific, from Chile to the East Indies. The crust is composed of several *plates* which move continuously against each other; where they are being created or destroyed, volcanoes and earthquakes occur and mid-ocean trenches and ridges form.

Lardarello Geothermal Field in Italy. Several hyperbolic cooling towers for the power plant are visible.

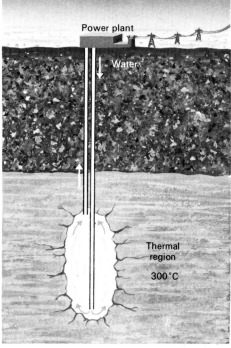

One possible method for extracting heat from dry rock several kilometres deep, as tested at Los Alamos, USA. The rock is first fractured vertically by hydrostatic pressure at the bottom of two bore holes. Sand is pumped in to hold the cavity open, then water circulated to a heat exchanger at the top.

body of water at 50–82°C, such as found in sedimentary deposits in Hungary, and its main use is for heating. Chemicals can also often be extracted from most fields.

It may be thought that there is little scope for geothermal energy in a country such as Britain. A report by ETSU, however, examines the fact that the temperature rises with depth in *any* hole drilled in the Earth's surface, but that parts of the UK, such as Cornwall, rest on 'hot rock' (usually granite) in which the temperature is higher than normal because of local heat sources. Water pumped into one drill hole and out of another could extract this heat. Indeed, it has been estimated that 520 cubic kilometres of underground rock a few hundred degrees hotter than the rock at the surface contains energy equal to that used worldwide in a year. ETSU suggests that the energy equivalent of 4 million tons of coal (4 *mtce* – million tons coal equivalent) a year could come from such sources in AD 2000, and Dr. William Bullerwell has pointed out to MPs that expensive equipment left in the North Sea when all its hydrocarbons have been recovered could be used for the purpose. Temperatures 4 km below the sea bed are as high as 160°C.

Sea power

Attractive though it is, direct solar energy is diffuse – spread thinly over the surface of the Earth – and needs to be concentrated before we can use it. (50 microgrammes of deuterium contain as much energy as a whole day's solar rays on a square metre of Earth's surface.) Except in space it is not always available where or when we need it. However, solar energy is responsible for a number of *indirect* forms of energy which can be harnessed, so that in effect we capture solar energy which arrived elsewhere. The Sun powers Earth's great weather engine,* providing over 3,000 Q per year for the purpose. Of this, one third is absorbed on passing through the atmosphere, while the rest – short wave radiation – heats the surface and is then returned to the atmosphere as long wave radiation and by the evaporation of water. Ultimately this heat is lost back into space at just about the same rate as it is received. As an energy source we are generally concerned here not so much with the heat as with the *mechanical* energy into which it is transformed; there is however at least one further use for the heating power of the Sun.

The idea of 'solar sea power', making use of the temperature differences between the warm surface layer and colder layers further down (especially in tropical

Principle of OTEC. One scientist has calculated that if *all* the world's electrical energy were produced by this method it would reduce the temperature of the tropical oceans by 1 °C, resulting in a reduction of tropical rainfall. . . .

* See *Air and Weather* by David A. Hardy (World's Work, 1977).

A Lockheed design for a giant OTEC plant.

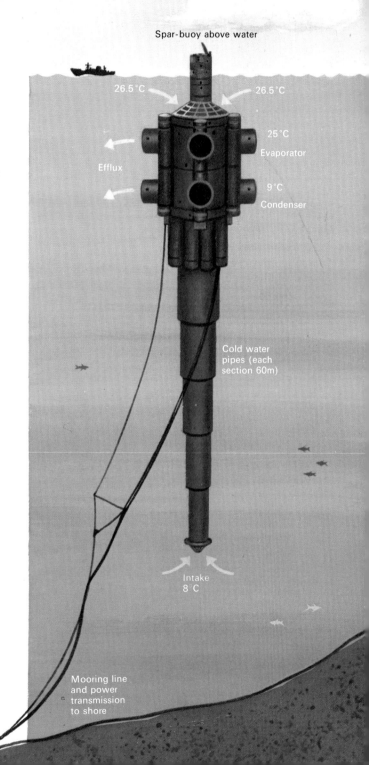

waters) was first put forward by a French physicist, Jacques d'Arsonval, in Paris in 1881. Now known as OTEC (ocean thermal energy conversion), the theory was put into practice in 1930 in a small way by Georges Claude, a student of d'Arsonval. His plant off the coast of Peru generated only 22 kW and was soon destroyed by heavy waves, and interest in the idea waned until 1964 when an American, J. Hilbert Anderson, became enthusiastic about the scheme – even though fossil fuels then seemed plentiful and cheap. In 1972 the National Science Foundation allotted $84,000 to his project. In 1975 this was increased to $3 million. Solar energy R & D then passed to ERDA, and in 1976 OTEC received $8 million. (If this sounds a lot, it should be compared with the $10 *billion* spent in 1975 on the fast breeder.)

There are various designs for OTEC, most having design features in common with that illustrated here. The 'working fluid', separate from the sea water and usually ammonia, is vaporised by the temperature differences – about 20°C – between surface 'heat reservoir', and cold water brought up by a pipe from the 'heat sink' 1,200 m below, driving generators. The cold water, as low as 5°C, then condenses the working fluid back to liquid for re-use. In addition to electricity, transmitted ashore by wire, such plants could make products such as hydrogen, oxygen and fresh water from the sea, or even ammonia and methanol from sea and surrounding air. All have energy applications.

Energy and the future

Gravity power

There are other uses for the energy of *moving* water, some of which Man has known for a great many years. Reference was made earlier to water-wheels, which use sun power indirectly – solar heat evaporates water from Earth's surface, forming clouds which later release that water as rain elsewhere. If that 'elsewhere' is high ground, gravity allows us to convert the kinetic energy of the flowing water into mechanical energy, and thence into work – whether that work is grinding corn or making electrical power. This principle was certainly used by the Greeks in 85 BC, and by the time of the scenario on page 58 there were probably 20,000 water-mills in use in Britain, with uses from grinding to operating hammers and bellows for the iron industry and pumping water. Today *hydro-electric* stations produce more electricity in the world than nuclear stations – though this may soon change.

The 'Pelton wheel', developed by L. A. Pelton in California in 1880, is the best

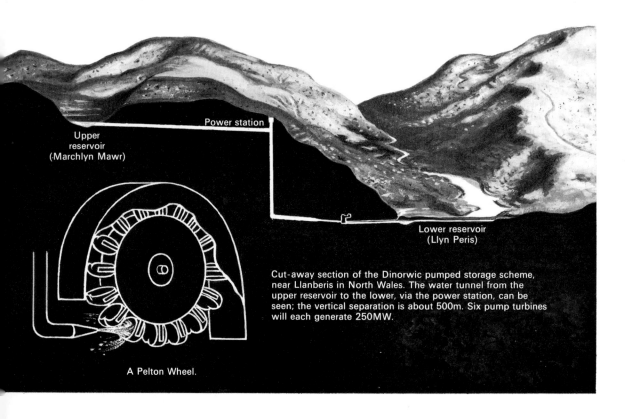

Cut-away section of the Dinorwic pumped storage scheme, near Llanberis in North Wales. The water tunnel from the upper reservoir to the lower, via the power station, can be seen; the vertical separation is about 500m. Six pump turbines will each generate 250MW.

A Pelton Wheel.

known *impulse turbine*, used where there is a low volume but a high 'head' of water – as from mountains – which jets from a nozzle into buckets around the rim of a wheel. These can deliver up to 60 MW. *Reaction turbines* are best used in fairly fast-moving rivers, which drive vaned wheels by flowing through an outer casing. These produce 500 MW, and could go up to 1,500 MW. A third type is the *axial flow* turbine, which resembles a ship's propellor encased in a wide tube. The efficiency of any of these turbines can exceed 90%. And to produce electricity and use it to power machines is more efficient than the old method of using the energy of the water directly through cogs and gears – attractive though an old water-mill looks in paintings.

Today, water is held back behind huge dams and confined in reservoirs in order to even out seasonal floods and droughts (which can still sometimes catch us out, as in 1976), to be used as required. Dams often have to be sited in areas of great scenic beauty, as for example in the US – where it is the fourth largest source of energy, Norway, Scotland and Wales, and ecological damage has to be avoided.

In order to be able to supply electricity at the time when the demand for it arises, *pumped storage* schemes were developed. The idea is simple: surplus electricity from off-peak periods (e.g. at night) is used to pump water to a reservoir at a high level, then the stored potential energy in the water is released to generate electricity at peak demand times, flowing into a lower reservoir. Electricity can also be produced at very short notice in the event of breakdown elsewhere. The classic example in Britain is the 360 MW station at Ffestiniog in North Wales (which has actually received a citation from the Welsh Tourist Board).

Although the Electricity Boards take pains to minimise the impact on the environment, any new scheme to flood a valley is likely to meet as much opposition as did plans to mine the biggest coal deposits found this century, in the Vale of Belvoir, Rutland; or to recover the 4,000 mtce of uranium ore recently discovered in a 6.5 × 1.5 km area of one of the Orkney Isles.

As there are not enough natural lakes around with convenient hills, an intriguing alternative use of pumped storage could be to make power stations drive compressors and pump compressed *air* into underground caves – which could be excavated with explosives – at off-peak times for storage. By supplying compressed air to the turbines instead of a large part of their own power being diverted for the purpose, their output could be more than doubled at peak times.

According to World Energy Conference (WEC) statistics, a world total capacity of 2,260 GW is theoretically available from hydro-electric power, with China, USSR and USA at the top of the league. It is interesting to note that South America, Africa and SE Asia, which are very short of coal and oil, have large amounts of water power available.

Energy and the future

1. Salter's ducks. A number of segments rock on a central tubular backbone, forming a 500m *string*, sideways-on to the waves. Pumps/turbines are housed in the inner section. Lanchester Polytechnic, Coventry, has tested a scale model on Loch Ness.
2. Cockerell's contouring rafts. Hydraulic motors and pumps between each raft convert the energy of wave motion into pressure in a fluid.
3. Air pressure ring buoy.
4. The Russell rectifier. Sea water is seen entering the upper reservoir through vertical non-return flaps, and (small drawing) leaving the lower, generating power on the way.

Moon power

Hydro-electric power is obviously available only in certain geographical locations. The energy of moving water is not limited to that flowing from high ground, though, anyone who has watched waves crashing against rocks or simply the sea's relentless ebb and flow on the beach must have felt its power. But could *this* be harnessed?

The alternate rising and falling of the sea's surface which we call *tides* are caused by the gravitational pull of the Sun and Moon, but especially the latter as it circles the Earth every 24 hours 50 minutes. The range in height is from 30 cm in the Mediterranean to 16 m in eastern Canada. A water-wheel making use of this twice-daily variation was in use in Britain (eg. at Wapping) as long ago as 1233; the usual system was to trap the incoming tide behind gates, then use out-flowing water to drive a mill when the gates were opened, after the tide had ebbed. The constantly-altering time of high tide was an obvious inconvenience. On a larger scale, the tidal flow can fill and empty a large reservoir, creating a difference in levels as in the hydro-electric system on the previous page, with or without pumped storage.

The first, and, apart from one Russian pilot project, only full-scale tidal installation is in the estuary of the Rance river in France, where the average tidal range is 8.4 m (peak 13.5 m). The storage basin has an area of 22 km² and the station has a maximum output of 240 MW. A number of schemes have been put forward over the years to build a *barrage* in the Severn Estuary, and although the CEGB has 'reluctantly abandoned' the project on grounds of cost* it has said that the scheme could contribute about 3% to the present total primary energy needs of the UK. Again, the environmental effects of such a barrage need to be examined carefully.

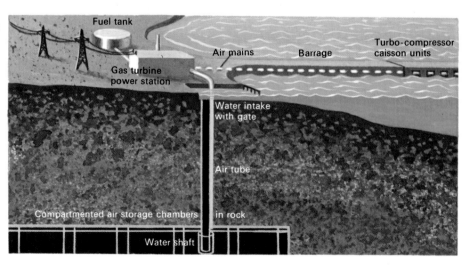

An unusual tidal/air storage scheme using gas turbines to generate electricity. Tidal turbines produce large quantities of compressed air which is stored in artificial underground caves and released to drive gas turbines at peak demand times. Some additional heat energy would be added at the time of generation.

*Estimated at £4 × 10⁹, and taking 20 years to build.

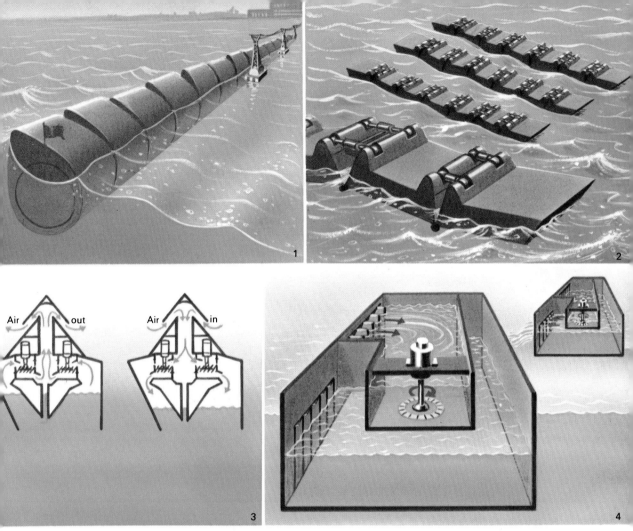

The sea has (at least) one further trick up its sleeve. Thanks very largely to the efforts of one man, Dr. Stephen Salter of Edinburgh University, Britain now has a £1 million *wave energy* R & D programme, based on four systems. This brings us into the province of 'wind energy' (hence solar, since winds are caused by the unequal heating of the atmosphere), but waves transfer energy with very high efficiency – waves arriving at one point may have originated in a storm a long way off.

From experiments in a tank with balsa wood, transistors and ballcocks, Salter improved the amount of energy taken out of artificial waves from 15% with an up-and-down movement to 60% with the to-and-fro. The eventual shape and movement of 'Salter's ducks', as they are now known, are shown in the diagram. Other wavepower devices – 'contouring rafts' – have been designed by Sir Christopher Cockerell's company, Wavepower Limited, and have been tested by British Hovercraft Corporation. Their aim is to produce simple, cheap, mass-produced units which can be installed and maintained a section at a time.

Other methods are the 'air pressure ring buoy', invented in Japan, (operating like an empty beer can with its open end under water) of National Engineering Laboratory, and the 'Russell wave rectifier' which combines wave power with high and low reservoirs, via valves.

Twelve wave power stations, each 80 km long, could supply half of Britain's current electricity needs. But it could be ten years before a 10 MW prototype is operating and 1996 or later before a 1 GW station is built.

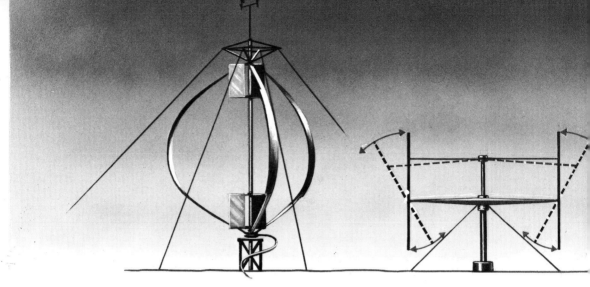

Two aerogenerator designs: *(left)* a Darrieus vertical axis and *(right)* Dr. Peter Musgrove's vertical axis machine. In the latter the blades' angle of attack changes with wind speed to achieve maximum efficiency.

Air power

From the energy of wind-driven waves, we can now turn to the wind itself. Windmills are another ancient method of obtaining power; those used by the Persians in AD 644 to drive pumps for irrigation had cloth sails and a vertical axle. The traditional 'picturesque' windmill with a tall tower and horizontal axle appeared in Europe at the beginning of the 14th century, and its numbers grew until, as we saw, steam made its takeover late in the 18th. In some countries, though, they reached a high standard of efficiency, especially when applied to the generation of electricity, as pioneered by Professor P. La Cour in Denmark. Wind-generated electricity was used up to and through World War II in Denmark, with over 480 MWh being obtained from 88 windmills in January 1944. The world's largest windmill was (not surprisingly) built in America. At Grandpa's Knob, Vermont in 1941 a 33.5 m-tall windmill produced 1.25 MW with the first synchronous electric generator. The introduction of rural electrification caused all but a few windmills to fall into disuse.

Wind is a notably capricious element. The UK's electrical needs could be met by tapping only a fraction of its available wind energy, but doing so economically is another matter. To compete with normal power stations, windmills or *aerogenerators* would have to be sited on hill-tops or round the coastline, where average wind speeds approach 32 kph. Even then a 50-m rotor would produce an average output of only some 350 kW. Modern windmills would not be as attractive as their predecessors; resembling more closely a large electricity pylon, a fast windmill could well also be noisy. They do have the advantage that the wind blows most when energy is most needed – in winter.

The power of a windmill depends on the area swept by its blades or rotor, its efficiency, and wind speed. The power available from the wind varies as the cube of the wind speed – which means that the power in a 100 kph wind is 1,000 times greater than one of 10 kph. Some sort of governor or method of shutting down in high winds is needed. Rotors are classified by their 'tip speed ratio', which is blade tip speed divided by wind speed. The theoretical possible efficiency – 59% – is unlikely to exceed 45% in practice. Most work best at 30–50 kph.

Modern technology (while sometimes going back to earlier designs, such as the Cretan canvas-sailed windmill – mainly for DIY home use) has evolved a variety of unusual designs, as shown here.

Comparison in size of a typical marsh drainage windmill with a design by ETSU/Servotec for a 1 MW aerogenerator and *(left)* a 400kV CEGB suspension tower (pylon). The Musgrove and Darrieus drawings are to approximately the same scale.

One solution to the visual and noise problem could be to site windmills offshore. A cluster of 400 windmills in an area 10 × 10 km could, according to some studies, give a total of 1 GW, with an average of 400 MW – comparable with a conventional power station, or 1.7% of our annual needs. As to storage during calm periods, among novel suggestions are to use excess electricity to electrolyse water, piping the hydrogen ashore to be stored for use as fuel (perhaps in fuel cells); and the use of a flywheel to even out fluctuations. A lig' fibre composite material for this purpose stores twenty times more energy than a solid steel flywheel, according to US work developed by Exeter University.

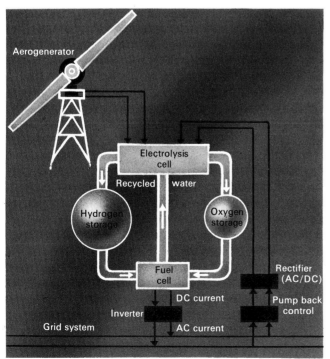

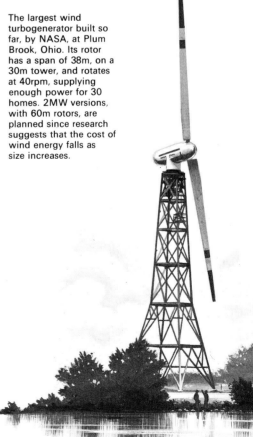

The largest wind turbogenerator built so far, by NASA, at Plum Brook, Ohio. Its rotor has a span of 38m, on a 30m tower, and rotates at 40rpm, supplying enough power for 30 homes. 2MW versions, with 60m rotors, are planned since research suggests that the cost of wind energy falls as size increases.

To store energy when the wind bows and release it continuously, the electricity generated could electrolyse water, the hydrogen and oxygen being stored, then fed steadily to a fuel cell. The direct current from this would be converted to AC and fed into the grid. Off-peak electricity from elsewhere could run the electrolysis cell in calm periods.

Plant power

We have seen that the fossil fuels on which we are presently so dependent were produced by solar energy from plants by photosynthesis. Of the massive income from the Sun only 1.2 Q per year is stored by plants in this way. (Since vegetation has existed for some 400 million years there should be total fossil fuel reserves of 480 million Q – but they only amount to 256 Q, which means an efficiency of conversion of two millions to one, or fossil fuel formation at 6×10^{-7} Q – equal to 16,000 tonnes of oil – a year.) It may be thought, therefore, that we can ignore plants and vegetation in our search for alternative energy sources. Certainly nobody would seriously suggest planting more forests so that we can burn the wood; yet it is estimated that 2,300 million tonnes of wood a year are used as fuel today, mainly by underdeveloped countries – a remarkably large amount, equal to about 15% of the world's 'normal' fuel consumption. This is a cause for some concern, since it could mean a great increase in carbon dioxide in the atmosphere; according to some sources, as much in a year as is produced by burning fossil fuels (5×10^9 tonnes per annum). Trees and plants 'mop up' this excess CO_2, some of the remainder being absorbed by the oceans, which contain 60 times as much as the air.

However, it is worth taking a look at the *water hyacinth*, a 'weed' which clogs rivers and canals in Africa, Asia and southern US. The water hyacinth can double in number every 8–10 days, responding especially to warm water (over 10°C). Because of this, it can be grown in the effluent from power stations, factories and sewage works; upon which it promptly purifies raw sewage and absorbs industrial pollutants, such as toxic metals, having a remarkable nutrient and mineral uptake. Because of its high protein and mineral content it can (except where toxic metals are absorbed), dried to below 15% moisture content, be used as a supplement in animal feed, and for the same reason it can be ground into an excellent compost and fertiliser.

As if this were not enough, water hyacinths can be used to produce *biogas* containing 60–80% methane – 374 litres of biogas per kilogram of dried plant, giving a fuel value two-thirds that of pure methane – by fermentation, when fed on sewage nutrients. 10^4 square metres can yield about one tonne of dry plant material per day, which can produce 220–440 cubic metres of methane – and the sludge can still be used as a fertiliser. Not bad for a weed!

ERDA, investigating 'biological solar energy conversion systems' (i.e. plants) rated them as 'most promising'. An ocean farm producing kelp in an area slightly larger than Texas, it is said, could supply the US with all its food and energy requirements. In Israel, Professor Mordhay Avron and Dr. Ami Ben-Amotz discovered that an unusual alga, *Dunaliella parva*, which grows in salty ponds along the Dead Sea, produced *glycerol* – up to 85% of its dry weight – to protect itself against lethal salt concentrations. Glycerol is a combustible alcohol which can be used to produce lubricants, pharmaceuticals, liquid soaps, cosmetics, and as a raw material for SNG and motor spirit. All *Dunaliella* needs is salt-water, CO_2 from the air, and sunlight; and it can be fed to animals. . . .

Water hyacinths were introduced into the USA by Japanese exhibitors at the Cotton Exposition in New Orleans in 1884; they had collected them from a river in Venezuela en route. NASA's Dr. Bill Wolverton is trying, not to kill of this beautiful but annoying and hardy weed, but to encourage its use — even urging that it be taken on an early Space Shuttle flight, since it could offer many advantages in the closed ecosystem of a space colony. Meanwhile, NASA's projected model farm could test a closed system on Earth.

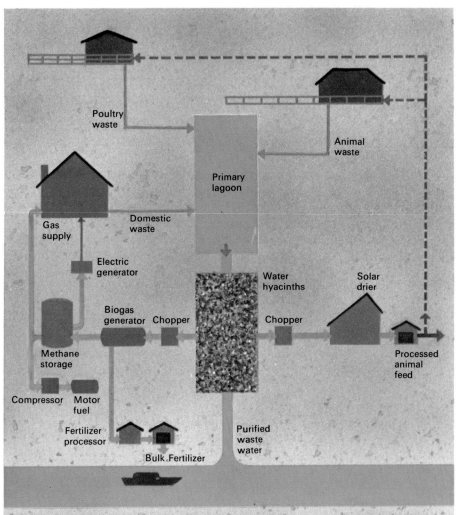

Energy and the future

And back to gas

Also living in very salty water in bright sunlight is a bacterium, *Halobacterium halobium*, which uses to produce energy not chlorophyll but a purple pigment which, by a strange quirk of nature, is very similar to the protein molecule in our eyes which converts incoming light into electrical signals to the brain. Light shining on its balloon-like membrane causes it to shoot out a proton, creating a

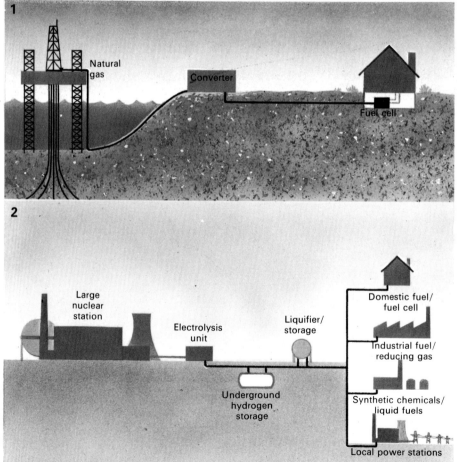

1. A fuel cell can be made to work on conventional hydrocarbon fuels converted into impure hydrogen. By obviating the need to convert fossil fuels first into heat, then mechanical energy, in order to produce electrical energy which then has to be transmitted by expensive and complicated methods (pages 76–77) to consumers, 'fuel cell power' would be much more efficient.
2. When fossil fuels run out and nuclear stations are built (needing to be sited away from cities and near to large amounts of cooling water to dissipate the 60–70% of low-grade heat they produce) new electrical transmission networks will be needed. Energy could be moved far more cheaply by underground (hydrogen) gas pipelines – often existing and already near coasts – than by overhead electric lines, while underground lines are even more costly.

'mini-electrical' potential which can be harnessed. According to Dr. Walther Stoeckenius of San Francisco, this organism could be used to generate electricity, desalinate sea water and produce important bio-chemicals.

A professor of chemistry at the University of North Carolina, David Whitten, has developed the first artificial chemical compound capable of splitting water into hydrogen and oxygen, using only the visible light from the Sun. Based on the metal *ruthenium*, Whitten's compound has a structure which resembles the active part of chlorophyll. Why it works is a mystery, even to Whitten and his colleagues, but the gases bubble off with high efficiency in the right conditions.

Before we get lost in more and more exotic solutions to the energy problem, let's come back to earth – rather literally. Mention has been made several times of the production of hydrogen (usually with oxygen, from H_2O) by various methods, and of methane and other gases. The 'hydrogen economy' appears attractive: hydrogen is a fuel – that is, a concentrated store of energy – and if it could replace petrol, say, it would have the advantage of being almost completely non-polluting, since burning hydrogen in air merely produces water, plus some oxides of nitrogen. It only stores about a third as much energy as natural gas, but being light has advantages when liquefied (though it requires a very low temperature, $-253°C$, and producing *this* uses energy). Hydrogen also explodes with 15 times less ignition energy than natural gas, so liquid hydrogen is not likely to be stored at garages like petrol! Even so, it is too good an energy source to ignore, and we have seen several ways to produce it. As for storage, substances such as *lanthanum-nickel* can soak it up like a sponge at $2\frac{1}{2}$ times atmospheric pressure at room temperature, forming a *hydride,* at nearly twice the density of liquid hydrogen. The hydrogen flows out as the hydride decomposes when the pressure is reduced. If the cost comes down, the hydrogen-powered car may become a reality, while Lockheed and NASA are studying a 6,000 kph liquid hydrogen-powered aircraft.

At any rate, when natural gas finally runs out there should be no excuse for the pipelines to stay empty. Hydrogen can be turned into methane by gasifying coal, with a nuclear reactor providing the energy if necessary. Or by combining with readily-available carbon it can be turned into liquid methanol, quite a good petrol-substitute. Alternatively (or in addition), as we saw earlier, gas can be produced from waste – garbage, or even sewage – either by pyrolysis or biological processes such as 'anaerobic digestion'. In the latter, bacteria 'feed' on animal wastes and decaying vegetable matter in the absence of air at a constant 30–35°C, producing biogas or methane. This idea is already in use on a small scale on farms and small-holdings (the exhausted material makes a fertiliser). The 20 million tonnes of garbage disposed of by Britain each year could give energy equal to 6 million tonnes of coal. In America the figure could be ten times this.

Energy and the future

The ideas keep coming

New ideas – and old ones revised – pop up all the time. Salt water features again in the 'solar pond' originated in Israel by Dr. Rudolph Bloch: heat absorbed from the Sun is stored in brine on the blackened bottom while clear water at the surface of the pond acts as an insulator, greatly reducing overnight heat loss. The water can approach 100°C, making this possibly the cheapest way of collecting solar energy on a large scale.

I have tried to outline all the major lines of research, those which seem most practical. There simply is not space here to describe every energy-producing or -saving invention; yet there is always a chance – however slim – that somewhere is *the* answer to all our future energy needs. Even if it does exist, the chances are that it could not be put into use on a large scale in time to have much effect before oil and gas, at least, run out. Some ideas seem to be strictly for do-it-yourselfers (yet even they can make a contribution. The National Centre for Alternative Technology near Machynlleth, which is a small self-supporting community, displays a variety of solar panels, windmills, biogas digesters and the like to the public and is well worth a visit if you are holidaying in North Wales); but our economy is such that it is clearly not practicable for every house to replace its rooftop TV aerial by an aerogenerator. Other schemes require a massive national effort on the scale of the Apollo programme. However, I cannot close without at least mentioning a few of the projects which did not quite 'fit in' earlier.

On the solar front, there is the 'solar eyeball' proposed by Derek Mash. This is a plastic sphere as big as a football with a gallium arsenide solar cell as its 'retina' on which the Sun's rays are concentrated by a lens. Thousands of these would float in tanks of water (perhaps in the desert), kept forever looking into the Sun by a system of magnetic pistons. Or Dr. Colin West's variation of the 150-year-old 'Stirling-cycle pump', which rocks in the Sun's rays as if by magic. Direct heat storage for domestic purposes etc. is now being investigated, using substances which absorb or surrender their latent heat as they change from solid to liquid or back. On the sea, a return to the days of sail is possible, but making use of modern aerodynamic principles. It is said that sail can compete with oil over long distances for some bulk cargoes. On the other hand, the ship is one of the few forms of transport that can use nuclear power directly (as submarines already do). As oil prices rise, such energy sources will soon become competitive.

Other developments aim to improve the efficiency of existing fuels. Professor Felix Weinberg has designed a new type of burner which can use much weaker mixtures of gas (eg. methane) in air than conventional burners, and at high temperatures and efficiency. Called the 'Swiss Roll' because of its shape and construction, it can also burn liquids or solids by incorporating a form of fluidised bed combustion, and could be used in, say, a locomotive. For power stations, *magneto-*

Even the intermittent movements of a tree can produce some 500W by means of a 'tree pump'. The cables and pulleys form a triangle and are connected to a chain-driven ratchet mechanism which rotates a shaft to produce mechanical/electrical power. (The vertical 'Tornado Tower' designed by Dr. James Yen of the Grumman Corporation, who estimates that its output could be 100 to 1,000 times that of conventional wind turbines, is a more practical use of wind energy.) At the opposite end of the energy scale, a magnetohydrodynamic generator can convert the energy in a hot, conductive gas directly into electricity by making its electrically-charged particles cut through the field of a powerful magnet.

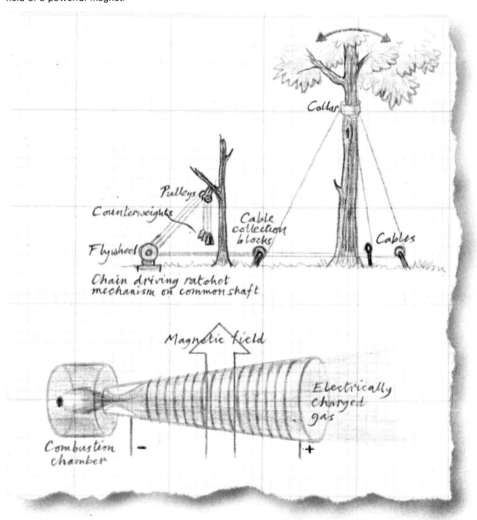

hydrodynamics (MHD) promises to boost conversion efficiency from the present 30% to 50–60%, by replacing the turbo-generator's present rotating parts with either a stream of liquid metal or a flow of hot, ionised gas passing through a magnetic field. *Superconductors* will also aid electricity generation and transmission; electrical resistance disappears at temperatures close to absolute zero ($-273°C$), giving theoretical efficiency of 100%. And if we ever get fusion power, the 'fusion torch' may convert all our garbage (even the wastes from 'out-dated' fission reactors!) back into raw materials. . . .

Energy and the future

Crisis — what crisis?

If one thing has become clear from all this, it is that there is absolutely *no* shortage of energy on Earth. But time is running out with the oil and gas, and one thing politicians must *not* do is to cut off funds for energy R & D. In Britain's case it is to be hoped that some of the revenue from North Sea oil will be used to develop alternative energy sources ready for when it runs out. An energy surplus in the 1980s could easily allow a false sense of security so that energy industries are not ready to meet later demands. Many people would prefer the use of renewable energy sources, free from temptation to terrorists and costing far less in R & D than the fast breeder programme. To be fair, though, there would probably be as much opposition to the construction of millions of huge, noisy windmills, or to the 1,500 km of wave-energy booms around our coasts needed to supply half of the UK's electrical needs, as to fast breeder stations.

A final factor to bear in mind is that all the energy locked up in the fuels we have examined, whether the chemical energy of fossil fuels or nuclear, ends up ultimately as *low-temperature heat* which cannot be used (even a heat pump only borrows it). En route it may have been converted into mechanical or electrical energy or into high-temperature heat or light; the point is that this 'new' heat is released constantly into our surroundings, and by the year 2000 could well begin to upset Earth's delicate climatic balance — as could the carbon dioxide from fossil fuels. Only renewable sources avoid this, since they use energy which arrived naturally and would otherwise simply have gone to waste.

It is only natural for the underdeveloped countries to want their share of the cake — the 'cake' in this case being the standard of living achieved by the industrialised nations. The capital cost of nuclear generating plant rules it out for most of these, which could give an ideal opportunity to develop and exploit various renewable systems. Of course, energy is not an isolated subject — it has to take its place alongside food production (itself an energy source), controlling the rapidly-growing populations who use that food and energy, resources in earth and sea, and so on.

Unpalatable though it may be to many of us, it seems that most industrialised nations are committed to the use of nuclear energy for the bulk of their future power. What is the outlook for other methods? Estimates vary greatly, and most are based on present-day technology and trends (whether it is desirable for us to continue to consume at an ever-increasing rate I will come to in a moment). As for technology, Dr. Chris Evans has pointed out that it almost always accelerates at a rate which outstrips predictions. Computers are a prime example. This may be a cause for optimism, though not for complacency. However, some conclusions can be drawn:

Only coal and some form of nuclear power are *capable* of supplying the total

Four special stamps issued in Britain in 1978 highlighted public awareness of the energy situation; they quickly sold out.

energy demanded by nations such as the USA and UK. The US has plumped for two types of LWR, while the UK favours the AGR – even though many feel this decision to be almost disastrous to our nuclear industry – and Canada CANDU. (It is worth mentioning here that Sir Fred Hoyle, in a thought-provoking and controversial book,* strongly favours the CANDU type of slow reactor over the fast breeder.)

Fast breeder reactors will be available by the year 2000 though and are, again, capable of supplying all electrical needs. Of the many other possible methods, predictions again differ, but the following picture emerges. An 'energy crunch' seems likely between 2000 and 2020: as oil and gas run out, the most acceptable and economically viable alternative will be coal-based synthetic fuels, especially as the price of uranium rises due to the depletion of lower-cost reserves, the fast breeder not having had time to become the mainstay of energy production. At the same time pressure will be put on the introduction of the fast breeder (unless public opinion or other factors dictate otherwise). To meet this demand, coal production will rise to perhaps 5 or 6 times today's, especially in the USA, by 2010. Its price will also rise, though, and the extraction of oil from shale and solar heating and cooling will become more attractive economically. Non-electrical energy costs can be expected to rise considerably faster than electrical – especially if fast breeders can be introduced at reasonable cost, in which case half of all energy used will probably be electrical. This will include a shift towards electricity for transport, either directly in vehicles or as electrolytic hydrogen fuel. Of the remaining, more speculative sources, fusion is usually left out of estimates because so many problems remain to be solved (though of course its potential in the long term is immense).

Solar electricity in one form or another is unlikely to provide more than a few percent of energy needs – but again is capable of much more, given vigorous R & D; the other most useful application of solar energy, apart from water and space heating, is the distillation of fresh water from saline. Geothermal and hydro-electric power is mainly limited by geographical locations, and unlikely to change the global picture greatly. Tidal, wave, wind, biomass/methane (most promising for farms) and waste/pyrolysis do not appear likely to contribute more than 2–10% each, even though the long-term potentials of wind and wave power and OTEC, for instance, are quite high – 25% or more. MHD and fuel cells remain somewhat unknown quantities, and hydrogen seems best suited to storage of wind-produced electricity, since if nuclear power stations prove capable of following changes in demand easily (which they cannot at present, being used to provide 'base load' while conventional stations are switched in at peak times) their power can be used directly. This situation could alter should coal prices double,

* Energy or Extinction? (Heinemann, 1977)

Energy and the future

Britain's energy consumption in 1975 was its lowest since 1969, thanks to the government's 'SAVE IT' campaign — which shows what *can* be done. Most house-owners could save money as well as energy if they increased existing insulation. Proper insulation could save the UK 80 million tonnes of coal a year by 2000 AD.

Department of Energy

or if the heat from nuclear reactors can be used to produce hydrogen efficiently *directly* from water.

If you feel strongly that our priorities have gone astray somewhere, NOW is the time to make sure that your government knows about it. There is, however, an even more important decision to be made, before it is made for us. The cost of all forms of energy is going to rise in real terms in the future, which means that growth rates will not be sustained at past or present levels. In other words, few of us will be able to go on 'living in the manner to which we have become accustomed'. This is likely to be a traumatic experience if we are unprepared; would it not be much more sensible to ask ourselves now whether we really want to continue our marriage to this demanding partner? Indeed, to agree now on an energy *target* instead of continuing to make demand predictions based on past trends? Once this is done, we can all help to force down that obdurate growth curve. What is needed is not consumption but *conservation*.

The first priority is to eliminate waste, which has insidiously become a part of our way of life. To take an example from the energy-producing methods discussed here: while pyrolysis can recover some energy from refuse (and incinerators linked to district heating schemes can be valuable), more energy could be saved by not making the unwanted objects in the first place! Disposable plastic articles, non-returnable bottles, unnecessary packaging and advertising material . . . we all know examples. Smaller cars should be encouraged, perhaps by a road tax based on the size of a vehicle's engine, and public transport improved. In the home, windows and doors should be draft-proofed and the former double-glazed where possible. Roof insulation is essential, and with cavity wall insulation could save 10% of our primary energy consumption. Thermostat settings for central heating only 1°C above what is really necessary can increase energy demand by 10%, and indeed with sensible clothing a cut from 23°C to 19°C can be perfectly acceptable. (We should not forget, though, to provide for increases where present levels are below reasonable standards, as in the case of many elderly people.)

All pipes and hot water cylinders should be insulated. At present perhaps 50% of domestic water heating is lost with waste water; heat can be recovered with a simple heat exchanger – equivalent to 2% of UK primary energy. Lights should be turned off when not needed. It is also worth noting that fluorescent tubes save over 70% for the same lighting levels. Most of the above points apply equally to industry and commerce, where sophisticated control systems could save 25% on heating. Properly designed new buildings could require only a tenth of the space heating of conventional houses, for about a 10% increase in cost; but such innovations need to be made now if they are to have any significant effect over the next 50 years.

Millions of tonnes of materials are burned, dumped, lost every year, when valuable substances could be recovered from them. Pulverised ash from power stations is used to make a light-weight concrete; but dirty sump oil worth £1 million is poured away each year in the UK alone, when it could be reclaimed or *usefully* burned. (Re-cycling schemes are coming into operation in many cities,* especially in the US.) All of the above is mere commonsense; there is no need to become an 'ecology freak', yet if everyone adopted these mainly simple and obvious ideas we could perhaps quite painlessly avoid that 'crunch'.

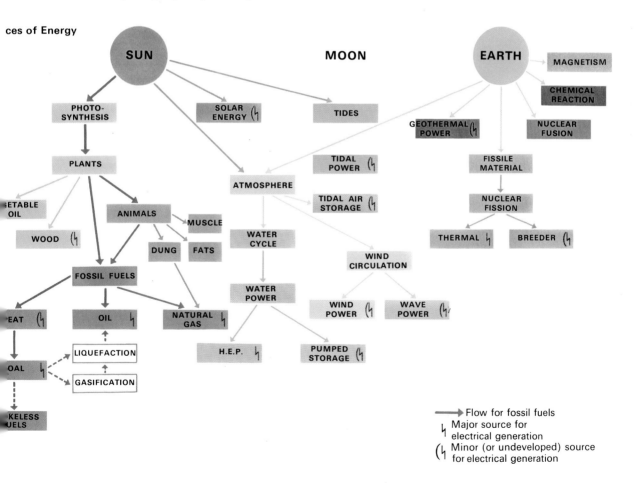

Sources of Energy

* Friends of the Earth and the Conservation Trust are happy to give details of local schemes.

Epilogue

'We could perhaps avoid that "crunch"'. Is that the best we can hope for? Is that a fitting way to close a book full of such marvels? In recent months the prophets of doom have been rampant, and it is no surprise if it seems that a mood of pessimism prevails. However, I for one cannot believe that calling a halt to industrial and economic growth, with the consequent lowering of living standards, frugality and general stagnation is either inevitable or acceptable to most people in the long run.

True, there is no excuse for waste, and it is time to put an end to a way of life based upon it. True, in the immediate future energy *will* become an expensive commodity, to be cherished and economised. And because present trends cannot be reversed overnight we *are* all likely to be poorer – by 10% or even 20% – by the year 2000 than we could have been, given continuing cheap oil, though still richer than we are now (but this assumes that we make the necessary adjustments in attitude as oil prices rise to as much as double today's by that crucial year. Otherwise the result will be a recession deeper than any yet experienced, with millions unemployed). True, the use of the private car for pleasure will have to be limited; and we have to take a decision on whether such personal transport is to continue to play a part in our future lives. If so, will it be powered by electricity, by methanol, or by hydrogen? But it is not Man's nature to regress, or to accept defeat.

We must ensure that this is no more than a temporary setback, and accept a period of relative privation in order to create a new base for growth; to employ Man's creativity and ingenuity (which has always come to the fore in times of adversity, such as during a war; witness the development of nuclear weapons, radar and the V2 – whose motor led directly to those in the Saturn rockets which took men to the Moon. We are now faced with a new kind of war, in which all men should be on the same side). We have long passed the stage where every family could be self-sufficient with its own plot of ground; only technology can pull us out of today's self-created morass. Nuclear power, because of decisions already taken, will take an increasing part in supplying our electrical energy, but recent events have turned the tide of public opinion in many countries, and the fast breeder reactor begins to seem less and less acceptable. Research on fusion power would be much more rewarding, and renewable energy sources become increasingly attractive – yet seem less likely because any financial encouragement for R & D may come too late to have an appreciable effect.

To me, personally, it seems obvious that *space* is the natural area for Man's necessary expansion, and outlet for his pioneering spirit. Leaving pure exploration aside, space offers huge advances not only in energy production but in industrial manufacturing processes (taking pollution away from Earth, and offering 'free' heat, cold, vacuum, zero gravity, etc., etc.), in education, communication, health, weather prediction and even climate control, and in discovering and shepherding resources on Earth and elsewhere in the Solar System. The USA has its Space Shuttle, which should herald a new era of growth in space; yet already shows signs of reducing rather than expanding its use. Only educated public opinion can reverse this negative trend. Even the UK could sponsor joint space projects with its European neighbours, on a scale much wider than it does today, broadening our options for growth – and indeed for survival.

Ironically, it was those first views of our Earth from space, relayed by Apollo over ten years ago, that awakened mankind's awareness of the vulnerable closed 'life support system' of our tiny planet as it circles an insignificant star, and aroused our environmental consciousness. There are of course alternative scenarios, not all pessimistic and some – even those based on conservation on a worldwide scale – quite attractive, and almost all requiring social change to a greater or lesser extent. Unfortunately, without the 'alternate universes' of science fiction there is not room for each person to choose his own future, and unless conscious decisions are made, and quickly, by both governments and individuals, the matter will be taken out of our hands. To me, it seems more fitting and sensible to build upon the technology which put men into space to provide for an exciting and secure future for our children instead of allowing it to go to waste. I also believe that, given the option, most people would find this a more acceptable and realistic solution, if only because one of the more immediate benefits of expansion into space would be the provision of employment for many thousands. The ultimate benefit would be to the entire human race; but it will not be achieved by apathy.

Selected bibliography

Energy *(especially statistics of resources, etc.)*
M. Kenward POTENTIAL ENERGY (Cambridge University Press, 1976)
G. Foley THE ENERGY QUESTION (Penguin, 1976)

Alternative energy sources etc.
G. Boyle LIVING ON THE SUN (Calder & Boyars, 1975)
G. Boyle & Peter Harper *et al.* RADICAL TECHNOLOGY (Wildwood House, 1976)
Various authors A SYMPOSIUM ON RENEWABLE SOURCES OF ENERGY (The Royal Society of Arts, 1976)
Workshop on Alternative Energy Strategies ENERGY – GLOBAL PROSPECTS 1985–2000 (McGraw-Hill (UK) Ltd, 1977)

Nuclear energy
NUCLEAR POWER AND THE ENVIRONMENT: 6th Report of Royal Commission on Environmental Pollution (HMSO, 1976)
F. Hoyle ENERGY OR EXTINCTION? (Heinemann, 1977)

General information

General information, booklets, visual aids lists etc. on the environment, resources, ecology, alternative technology and allied subjects may be obtained from the following organisations:
The Conservation Trust, 246 London Road, Earley, Reading, RG6 1AJ.
Friends of the Earth, 9 Poland Street, London, W1V 3DG.
Centre for Alternative Technology, Machynlleth, Powys, N. Wales.
E.G.I.S. Environmental Information Service, North Lodge, Elswick Road Cemetery, Newcastle-upon-Tyne, NE4 8DL.
Pictorial Charts Education Trust, 27 Kirchen Road, London, W13 0UD.
The L-5 Society (solar satellite power etc.) West European Branch,
45 Wedgewood Drive, Lilliput, Poole, Dorset, BH14 8ES.
Information on gas, coal, electricity etc. is usually available from the specific Board or Corporation involved; similarly with the oil companies (especially British Petroleum).

Acknowledgements

The author's thanks go to the following organisations, individuals and sources for supplying information and assistance in what proved quite a mammoth but enjoyable task; to his helpers Ruth and Mary for their invaluable work on the Index; and to Tony Osman of *The Sunday Times* for his help in reading the proofs and for his helpful comments.

The chart on page 103 has been adapted by the author from a chart by Peter S. Berry of the Conservation Trust, with his kind permission.

David Baker
C. J. Beavor, Department of Industry, Birmingham
Berkeley Nuclear Laboratories
British Gas
British Petroleum Company Ltd
Central Electricity Generating Board
The Conservation Trust
Department of Energy
Alan Dobson (*Tomorrow's World*)
Electrical Research Association
ESB Incorporated (USA)
ETSU, Harwell
David Fishlock (*Financial Times*)
Friends of the Earth
Institute of Petroleum
International Solar Energy Society
JEOL (UK) Ltd
Dale Kornfeld (NASA)
Lanchester Polytechnic, Coventry
The L-5 Society
Dr. Anthony Michaelis
Dr. Peter Musgrove, University of Reading
National Centre for Alternative Technology
National Coal Board
National Space Institute (USA)
New Scientist
Royal Society of Arts
Dr. Stephen Salter, University of Edinburgh
Scientific American
Solar Energy Consortium
Technology Review
Underwater Journal
UK Atomic Energy Authority
Wavepower Ltd
Weizmann Institute of Science (Israel)

Index

Italic figures indicate that an illustration appears.

Acceleration 17 18
Advanced gas-cooled reactor (AGR) 70 *71* 74
Aerogenerator 92 *92 93*
Alpha rays 45 *45* 46 *46*
Ampere 35 *35*
Ampère, André Marie 35 *36* 37
Anaerobic digestion 96
Anderson, J. Hilbert 87
Anode 42 *42 43* 82
Anthracite 54 *55* 57
Anticline *58*
Antielectron 50
Antineutrino 45 48
Apollo 10 *11* 14 18 31
Aristotle 10 *11*
Asphalt 64
Atom 23–31 *24 25 26* 27 *29 30 41* 41–50 70
Atomic bomb *47* 49
Atomic number 27
Atomic Theory 23
Atomic weight 24 25 28
Avogadro, Amedeo *25*
Avron, Professor Mordhay 94

Barrel(oil)62 70
Battery 33 *33* 82
Becquerel, Henri 44 45
Ben-Amotz, Dr Ami 94
Berzelius, Jöns 24 *24*
Beta rays 45 *45* 46
Biogas 94 97
Bitumen 64 *65*
Black, Joseph 21
Bloch, Dr Rudolph 98
Bohr, Niels *26* 27 28 41 43
Boyle, Robert *16* 20 23
British Nuclear Fuels Ltd., 73 *74*
British Thermal Unit (Btu) 18 70
Brown, Robert 25
Brownian Motion 25
Butane 59 60

Caloric 21–22
Calorie 18 19 30 70
Calorimeter 30
Calorimetry 20
CANDU 74 101
Carbohydrates 54 *55* 60

Carbon 24 *26 27 29* 49 54 *55* 60
Carlisle, Sir Anthony 31 33
Carnot, Lazare 14
Cathode 42 *42 43* 82
Cathode rays 43 44
Cathode ray tube 42 *42 43 43*
Cavendish, Henry 24
Celsius scale 20
Chadwick, James 27 48
Chain reaction *48* 49
Char 58
Charles, Jacques 21
Charm 48
Chlorophyll 54 *55*
Claude, Georges 87
Clean Air Act 57
Cloud chamber *46*
Coal 53–57
Coal, bituminous 54 *55* 57
  by-products 57
  cannel 54 *55*
  deposits 57
  formation 54 *55*
Cockerell, Sir Christopher 91
Cockerell's contouring rafts *91*
Coke 57 *57* 58
Compound 23 24 44
Compressed air 89 *90*
Compressed gas 22 *23*
Compton, Arthur H. 40
Conduction *15* 17
Conductors 34 36
Conservation of Energy, Law of 13 14 44
Conservation of Matter, Law of 44
Convection *15* 17
Cosmic rays 50
Coulomb 35 *35*
Coulomb, Charles Augustin 35
Covalent bonding 30 *30*
Critical mass 49
Crookes, Sir William 42
Crude oil 62 64 *64* 69
Curie, Marie and Pierre 44–45

Dalton, John 23 24 *24* 25 28
Darrieus vertical axis *92*
Davy, Humphrey 22 33
Decay, alpha, beta 44
Democritus 10 23

Derrick 66 *67*
Deuterium *27* 50 *50* 74 75 86
Diesel, Rudolf 53
Discharge tube 38 42
Dounreay 74
Drake, Edwin L. 65
Du Fay, Charles 31
Dunaliella Parva 94–95
Dynamo *37* 38
Dyne (dyn) 17 18

Edison, Thomas 36
Einstein, Albert 40 41 44 48
Electricity, static 31 32 *32* 34 42
   transmission and distribution 76 77
Electrode *33* 42 *82*
Electrolysis 31 33 *33* 93 *93*
Electrolyte *33 82*
Electromagnetic induction *36* 37 76
Electromagnetic spectrum *38* 39–40
Electromotive force (emf) *35*
Electroplating 33
Electrostatic induction 40
Electron *26* 27–30 34–35 40–41 *41 43* 49
Electron microscope 25 *26*
Element 23–24 44
Endothermic reaction 25
Energy, chemical 25
   heat 14 20–22
   internal 17 *19* 20
   kinetic *12* 13 14
   mechanical 18
   potential *12* 14 15
   world production and consumption *52*
Erg 17
Exothermic reaction 25

Fahrenheit, Daniel 20 21
Fallout 50
Faraday, Michael *33 36* 37 38
Fermi, Enrico 48 49 *49*
Fluidised bed combustion 58
Foot-pound 17
Fossil fuels 10 53 54 *55 70* 94
Fractional distillation *68* 69
Franklin, Benjamin 31 32
Friction 14 *15 19* 38
Fuel cell 82 *82* 93 96 97

Galileo 10 *11* 13 14
Galvani, Luigi 32 33 40
Galvanometer 37
Gamma rays *38* 40 45 *45 46* 50
Gas deposits *58* 59
Gasification 58
Gay-Lussac, Joseph 21
Gell-Mann, Murray 48
Generator *36* 37 76–77
Geothermal energy 84 *84* 85 *85*
Geysers 84
Gilbert, Dr William 34
Glycerol 94 95
Goldstein, Eugen 42
Gramme, Z.T. 76
Grid, national 76 77 *77* 93
Grimaldi, Francesco 39
Guericke, Otto von 31

Hahn, Otto 48
Half-life 46
Hallwachs, Wilhelm 40
Halobacterium halobium 96
Heat pump system 80
Heliostat 80
Helium *27* 27 50 *50*
Helmholtz, Hermann von 13
Henry, Joseph 37
Herschel, Sir William 39
Hertz, Heinrich 40
Hooke, Robert 20
Horsepower 17
Hoyle, Sir Fred 101
Huygens, Christian 13 *16* 39
Hydrates 65
Hydroelectric power 88–89
Hydrogen 24 *24* 27 *27* 28 *30* 49 *75*
   93 *93* 97
Hydrogen bomb 49 50 *50*

Infra-red *38* 39 80
Insulators 34
Internal combustion engine 53
Ion 30 *46 51 82*
Ionic bonding *30* 30
Isotope 27 *27* 46 74

Joule 18 19 70
Joule, James Prescott 18 19

Kelvin 20
Kerogen 69
Kerosene *64* 65 *65*

La Cour, Professor P. 92
Lanthanum-nickel 96
Latent heat 21 22
Lattice 25 *26 30*
Lavoisier, Antoine 24
Leibniz, Gottfried Wilhelm 13
Lenard, Philipp 40
Leyden jar 32 *32*
Light Water Reactor 71
Lignite 54 *55* 57
Liquefaction 57
Liquefied natural gas (LNG) 60 63 *63*
Liquefied petroleum gas (LPG) 60 62
Lithium *27* 75
Lurgi gasifier *56*

Magnetohydrodynamics (MDH) 98 *99* 101
Magnox reactor 70 *71 74*
Manhattan project 49
Mash, Derek 98
Mass 13
Mass number 27
Maxwell, James Clerk 39
Mayer, Robert 22 23
Megaton 49
Mendeléev, Dmitri 28
Methane 59 60 64 65 94 *95* 97
Methanol 62 63 87
Microwaves *39 82 83*
Moderator 49 *49* 70
Molecular weight 28 69
Molecules *15* 17 *19* 23 25 *26* 30 *30*
Motion, Laws of 14
Musgrove, Dr Peter *92*

Naptha *57 64* 65 *65*
Natural gas 58–63 *61 70*
Natural gas liquids (NGL) 62
Neutron 27 *27* 45 48 *48* 49 50 70 72
Newton 14 18
Newton, Sir Isaac 13 14 *16* 39
Nicholson, William 31 33
Nuclear energy 70–75 101
Nuclear fission 48 *48* 50 70
 fuel cycles *74*

fusion 49 50 75 *75*
 waste disposal 73
Nucleus *26* 27 30 45 49

Octane chains *69*
Oersted, Professor Hans Christian 37
Ohm 34 *35*
Ohm, George Simon 34
Oil, by-products *65*
 deposits *58* 59 *59* 65
 drilling for 65 66 *67*
Oil-sands 69
Oil-shale 69 *70*
Orbital *29*
OTEC *86* 87 *87*
Otto, N. A. 53
Oxidation 25 54
Oxygen 24 *24* 25 28 *30* 31 96

Paraffin 65 *68*
Paraffinic hydrocarbons 60
Pelton wheel 88 *88*
Pentane 59
Periodic table 28 29 *29* 44 45
Perrin, Jean 42
Petroleum 64 *64* 65 *68*
Phlogiston theory 24
Photoelectric effect 40 41
Photoelectrons 40
Photons 40 41
Photosynthesis 54 *55* 70
Pitch 64
Planck, Max 40
Plücker, Julius 42 *42*
Plutonium 49 50 72 73 *74*
Positron 50
Priestley, Joseph 24
Primary fuels 53
Production platform 66 *67*
Propane 59 60 62
Proton 27 34
Proton accelerator 48
Pumped storage scheme *88* 89
Pyrolysis 57 58 97

Quantum 40 41
Quark 48

Radar *39*
Radiant heat 39

Radiation *15* 17 40 50
Radio waves *39*
Radioactivity 44–50 73
Radioisotope 50
Radium 45 *45*
Reactor 70 *71 72* 72–75
Rectenna *83*
Refinery *68* 69
Relativity, theory of 44
Resistance, electrical 34 *35* 38
Resistors 36
Ritter, Johann Wilhelm 39
Röntgen, Wilhelm 42
Rotor 92 *93*
Russell rectifier 91 *91*
Rutherford, Lord Ernest 27 43 45 *45* 46

Salter, Dr Stephen 91
Salter's ducks 90 91 *91*
Schweigger, Johann *36* 37
Scott, David 10 *11*
Seeps (oil) 59 62
Seismogram 66
Smeaton, John 53
Smokeless fuels 57 58
Solar cell 82 98
    eyeball 98
    pond 98
    power station *83*
    water heating 80 *80* 81 *81* 98
Sommerfeld, Arnold 27
Specific gravity, degree of 69
Specific heat 21 *21* 22
Spectrum 39
Stahl, Georg Ernst 24
Stirling-cycle pump 98
Stoeckenius, Dr Walther 96 97
Strangeness 48
Strassman, Fritz 48
Stratigraphic trap *59*
Substitute natural gas (SNG) 58 63
Superconductors 99
Syncrude 57 69
Synthesis gas *56* 57

Tar-sands 69 *70*
Thermal density 62 *62*
    energy 62
Thermodynamics 19–20 21

Thompson, Sir J. J. 42 *43*
Tidal energy 90 *90*
Tokamak 75 *75*
Tornado tower *99*
Town gas 58
Transformation series 46
Tree pump *99*
Tritium 27 50 *50* 75
Turbine, axial flow 89
    gas 58 *90*
    impulse 89
    reaction 89
Turbogenerator *93*

Uranium 44 46 48–50, *49 50* 70 72 73 *74* 89
Ultra-violet *38* 39 40

Valence 25
Van der Waals forces *26* 30
Villard, P. 45
Vis viva 13 14
Volta, Alessandro 33 *33*
Voltage 77 *77*
Voltaic pile 33 *33*
Volts 34 35 *35*

Water hyacinth 94 *95*
Watson, Sir William 32
Watt 17 18 35 *35*
Watt, James *16* 17 35 53
Watt hour 19
Wavelengths 40 *41*
Wave power 91 *91*
Weinberg, Professor Felix 98
West, Dr Colin 98
Whitten, David 97
Wien, Wilhelm 42
Wilson cloud chamber 46 *46*
Wimshurst machine *32*
Wind energy 92–93
Windmills 92 93 *93*
Windscale 70 73
Wolverton, Dr Bill *95*
World energy sources *52 103*

X-rays *38* 40 42 *43* 45

Yen, Dr James *99*
Young, Dr Thomas 17 39